1 MONTH OF
FREE
READING

at

www.ForgottenBooks.com

By purchasing this book you are eligible for one month membership to ForgottenBooks.com, giving you unlimited access to our entire collection of over 1,000,000 titles via our web site and mobile apps.

To claim your free month visit:

www.forgottenbooks.com/free845019

ISBN 978-0-365-28281-5
PIBN 10845019

This book is a reproduction of an important historical work. Forgotten Books uses
state-of-the-art technology to digitally reconstruct the work, preserving the original format
whilst repairing imperfections present in the aged copy. In rare cases, an imperfection in
the original, such as a blemish or missing page, may be replicated in our edition. We do,
however, repair the vast majority of imperfections successfully; any imperfections that
remain are intentionally left to preserve the state of such historical works.

THE

WESTERN WORLD.

PICTURESQUE SKETCHES OF
NATURE AND NATURAL HISTORY IN NORTHERN
AND CENTRAL AMERICA.

BY

WILLIAM H. G. KINGSTON.

WITH EIGHTY-SIX ENGRAVINGS

London:
T. NELSON AND SONS, PATERNOSTER ROW.
EDINBURGH; AND NEW YORK.

1884.

Preface.

— ❖ —

IN the following pages I have endeavoured to give, in a series of picturesque sketches, a general view of the natural history as well as of the physical appearance of North America.

I have first described the features of the country; then its vegetation; and next the wild men and the brute creatures which inhabit it. However, I have not been bound by any strict rule in that respect, as my object has been to produce a work calculated to interest the family circle rather than one of scientific pretensions. I have endeavoured to impart, in an attractive manner, information about its physical geography, mineral riches, vegetable productions, and the appearance and customs of the human beings inhabiting it. But the chief portion of the work is devoted to accounts of the brute creation, from the huge stag and buffalo to the minute humming-bird and persevering termites,—introduced not in a formal way, but as they appear to the naturalist-explorer, to the traveller in search of adventures, or to the sportsman; with

descriptions of their mode of life, and of how they are found,
hunted, or trapped. I have described in the same way some
of the most remarkable trees and plants; and from the accounts I have given I trust that a knowledge may be obtained
of the way they are _____ d how their produce is
prepared and employed pe that, with the aid of
the numerous illustratio :k, a correct idea will be
gained of the wilder and more romantic portions of the great
Western World.

WILLIAM H. G. KINGSTON.

Contents.

NORTH AMERICA.

CHAPTER VI.
RODENTS.

CHAPTER VII.
CARNIVORA.

CHAPTER VIII.
THE FEATHERED TRIBES OF NORTH AMERICA.

CHAPTER IX.
REPTILES.

MEXICO AND CENTRAL AMERICA.

CHAPTER I.
MEXICO.

CHAPTER II.
CENTRAL AMERICA.

CHAPTER III.
RUINS OF CENTRAL AMERICA.

List of Illustrations

THE WESTERN WORLD.

NORTH AMERICA.

CHAPTER I.

INTRODUCTORY.—PHYSICAL FEATURES OF NORTH AMERICA.

THE continent of America, if the stony records of the Past are read aright, claims to be the oldest instead of the newest portion of the globe.* Bowing to this opinion of geologists till they see cause to express a different one, we will, in consequence, commence our survey of the world and its inhabitants with the Western Hemisphere. From the multitude of objects which crowd upon us, we can examine only a few of the most interesting minutely ; at others we can merely give a cursory glance ; while many we must pass by altogether,—our object being to obtain a general and retainable knowledge of the physical features of the Earth, the vegetation which clothes its surface, the races of men who

---

* According to some geologists, Labrador was the first part of our globe's surface to become dry land.

inhabit it, and the tribes of the brute creation found in its forests and waters, on its plains and mountains.

As we go along, we will stop now and then to pick up scraps of information about its geology, and the architectural antiquities found on it; as the first will assist in giving us an insight into the former conditions of extinct animals, and the latter may teach us something of the past history of the human tribes now wandering as savages in regions once inhabited by civilized men.

Still, the study of Natural History and the geographical range of animals is the primary object we have in view.

Though the best-known portions of the Polar Regions are more nearly connected with North America than with Europe or Asia, we propose to leave them to be fully described in another work. It is impossible, in the present volume, to embrace more than the continental parts of the Western World.

Looking down on the continent of North America, which we will first visit, we observe its triangular shape : the apex, the southern end of Mexico ; the base, the Arctic shore ; the sides, especially the eastern, deeply indented, first by Hudson Bay, which pierces through more than a third of the continent, then by the Gulf of St. Lawrence, and further south by Chesapeake Bay and the Bay of Fundy. On the western coast, the Gulf of California runs 800 miles up its side, with the Rio Colorado falling into it ; and further north are the Straits of Juan da Fuca, between Vancouver's Island and the mainland, north of which are numerous archipelagoes and inlets extending round the great peninsula of Yukon to Kotzebue Sound.

Parallel with either coast we shall see two great mountain systems—that called the Appalachian, including the chain of tho Alleghanies, on the east, and the famed Rocky Mountains on the west—running from north to south through the continent.

We shall easily recollect the great water-system of North America if we consider it to be represented by an irregular cross, of which the Mississippi with its affluents forms the stem ; Lake Superior and the River St. Lawrence, including the intermediate lakes, the eastern arm; the Lake of the Woods and its neighbours, Lake Winnepeg and the Saskatchewan, the western arm ; and the northern lakes of Athabasca, the Great Slave Lake, and the Mackenzie River, the upper part of tho cross. If we observe also a wide level region which runs north and south between the Arctic Ocean and the Gulf of Mexico, bounded on either side by the two lofty mountain ranges already mentioned, we shall have a tolerably correct notion of the chief physical features of the North American continent.

Arriving at the northern end, we shall find it reaching some four degrees north of the Polar Circle, though some of its headlands extend still further into the icy sea. Beyond it stretches away to an unknown distance towards the Pole a dense archipelago of large islands, the narrow channels between them bridged over in winter by massive sheets of ice, affording an easy passage to the reindeer, musk-oxen, and other animals which migrate southward during the colder portion of the Arctic winter.

NORTHERN REGION.

With that end of America will ever be associated the names of Sir John Franklin and his gallant companions, who perished in their search of the North-west Passage; as well as those of other more fortunate successors, especially of Captains M'Clure and Collinson of the British navy, to the first of whom is due the honour of leading an expedition from west to east along that icy shore ; while Captain Collinson took his ship, the *Enterprise*, up to Cambridge Bay, Victoria Land, further east than any ship had before reached from the west —namely, 105° west—and succeeded in extricating her from amid the ice and bringing her home in safety.   Captain M'Clure, not so fortunate in one respect, was compelled to leave his ship frozen up.   The two expeditions, while proving the existence of a channel, at the same time showed its uselessness as a means of passing from the Atlantic to the Pacific, as, except in most extraordinary seasons, it remains blocked up all the year by ice.

The northern end of the American continent is a region of mountains, lakes, and rivers.   Several expeditions have been undertaken through it,—the first to ascertain the coast-line, by Mackenzie, Franklin, Richardson, Back, and others, and latterly by Dr. Rae ; and also by Sir John ,Richardson, who left the comforts of England to convey assistance to his long-missing former companions, though unhappily without avail.   These journeys, through vast barren districts, among rugged hills, marshes, lakes, and rivers, in the severest of climates, exhibit in the explorers an amount of courage, endurance, and perseverance never surpassed.   In the course of the rivers occur many dangerous falls, rapids, and cataracts, amid rocks and huge boulders, between which the voyagers' frail barks make their way, running a fearful risk every instant of being dashed to pieces.   Not a tree rears its head in the wild and savage landscape, the vegetation consisting chiefly of lichens and mosses.   Among the former the tripe de roche is the most capable of supporting life. Here winter reigns with stern rigour for ten months in the year; and even in summer biting blasts, hail-storms, and rain frequently occur. Yet in this inhospitable region numerous herds of reindeer, musk-oxen, and other mammalia find subsistence during the brief summer, as do partridge and numerous birds of various species.

Here the Esquimau lives in his skin-tent during the warmer months, and in his snow-hut in winter, existing on the seals which he catches with his harpoon, the whales occasionally cast on shore, and the bears, deer, and smaller animals he entraps.

The numerous rivers flowing from the mountain-ridges mostly make their way northward. The Mackenzie, the largest and most western, rising in the Great Slave and Great Bear Lakes, falls, after a course of many hundred miles, into the Polar Sea. The Coppermine River, rising in Point Lake, makes its course in the same direction; while eastward, the great Fish or Back River, flowing from the same lake as the first mentioned stream, reaches the ocean many hundred miles away from it, at the lower extremity of Bathurst Islet. It runs rapidly in a tortuous course of 530 geographical miles through an iron-ribbed country, without a single tree on the whole line of its banks, expanding here and there into five large lakes, and broken by thirty-three falls, cascades, and rapids ere it reaches the Polar Sea. Not far from its mouth rises the barren rocky height of Cape Beaufort.

It was down this stream that Captain Back, the Arctic explorer, made his way, but was compelled to return on account of the inclemency of the weather and the difficulty of finding fuel; the only vegetation which he could discover

being fern and moss, which was so wet that it would not
burn, while he was almost without fire, or any means of
obtaining warmth, his men sinking knee-deep as they proceeded on shore in the soft slush and snow, which benumbed
their limbs and dispirited them in the extreme.   Through
this country the unhappy remnant of the Franklin expedition,
many years later, perished in their attempt to reach the
Hudson Bay Company's territory.   Here, in winter, the thermometer sinks 70° below zero.   Even within his hut, when he
had succeeded in lighting a fire, Back could not get it higher
than 12° below zero.   Ink and paint froze.   The sextant cases,
and boxes of seasoned wood — principally fir — all split ; the
skin of the hands became dried, cracked, and opened into un-
sightly and smarting gashes; and on one occasion, after washing
his hands and face within three feet of the fire, his hair was
actually clotted with ice before he had time to dry it.   The
hunters described the sensation of handling their guns as
similar to that of touching red-hot iron; and so excessive was
the pain, that they were obliged to wrap thongs of leather
round the triggers to prevent their fingers coming in contact
with the steel.   Numbers of the Indian inhabitants of the
country perish from cold and hunger every year—indeed, it
seems wonderful that human beings should attempt to live in
such a country; yet much further north, the hardy Esquimau,
subsisting on whale's blubber and seal's flesh, contrives to
support life in tolerable comfort.

To the south of the Arctic Circle stunted fir-trees begin to
appear, and at length grow so thickly, that it is with difficulty a passage can be made amid them.   Frequently the
explorer has to clamber over fallen trees, through rivulets,
bogs, and swamps, till often the difficulties in the way

appear insurmountable to all but the boldest and the most persevering.

MOUNTAINS.

On the western side of the continent rises gradually from the Polar regions the mighty chain which runs throughout its whole length—a distance of altogether 10,000 miles. The northern portion, known as the Rocky Mountains, runs for 3000 miles, in two parallel chains, to the plains of Mexico, flanked by two other parallel ranges on the west,—the most northern of which are the Sea Alps of the north-west coast, and on the southern, the mountains of California. At the north-western end of the Sea Alps rises the lofty mountain of Mount Elias, 17,000 feet in height—the highest mountain in North America — not far from Behring Bay; while another range, the Chippewayan, stretches eastward, culminating in Mount Brown, 16,000 feet in height, and gradually diminishing, till it sinks into insignificance towards the Arctic Circle. Point Barrow is the most northern point of America on the western side. It consists of a long narrow spit, composed of gravel and loose sand, which the pressure of the ice has forced up into numerous masses, having the appearance of rocks. From this point eastward to the mouth of the Mackenzie River the coast declines a little south of east. The various mountain ranges existing on the eastern side of the continent, including the chain of the Alleghanies, form what is called the Appalachian system. It consists of numerous parallel chains, some of which form detached ridges, the whole running from the north-east to the south-west, and it extends about 1200 miles in length—from Maine to Alabama. Besides the Alleghany Mountains in the western part of Virginia and the central parts of Pennsylvania, it embraces the Catskill Mountains in the State of New York, the Green Mountains in the State of Vermont, the highlands eastward of the Hudson River, and the White Mountains in New Hampshire. Mount Washington, which rises to an elevation of 6634 feet out of the last-named range, is the highest peak of the whole system. To the north of the St. Lawrence the lofty range of the Wotchish Mountains extends towards the coast of Labrador; while the whole region west and north of that river and the great Canadian lakes is of considerable length, the best-known range being that which contains the Lacloche Mountains, which appear to the north of Lake Huron, and extend towards the Ottawa River. These two great ranges of mountains divide the North American continent into three portions.

GREAT RIVERS.

The rivers which rise on the eastern side of the Appalachian range run into the Atlantic; those which rise west of the Rocky Mountains empty themselves into the Pacific; while the mighty streams which flow between the two, pass through the great basin of the Mississippi, and swell the waters of that mother of rivers. The great valley of the Mississippi, indeed, drains a surface greater than that of any other river on the globe, with the exception perhaps of the Amazon. The Missouri, even before it reaches it, runs a course of 1300 miles, while the Mississippi itself, before its confluence with the Missouri, has already passed over a distance of 1200 miles; thence to its mouth its course is upwards of 1200 miles more. The Arkansas, which flows into it, is 2000 miles long, and the Red River of the south 1500 miles in length; while the Ohio, to its junction with the Mississippi, is nearly 1000 miles long.

North America may be said to contain four great valleys
—that of the Mississippi, running north and south ; that of
the St. Lawrence, from the south-west to the north-east ; that
of the Saskatchewan, extending from the Rocky Mountains
below Mount Brown to Lake Superior ; that of the Mackenzie, from the Great Slave Lake to the Arctic Ocean.
Although a large portion of the eastern side of the continent is
densely wooded, there are towards the west, extending from
the Gulf of Mexico to the Arctic Ocean, vast plains.   In the
south they are treeless and barren in the extreme ; while
advancing northward they are covered with rich grasses,
which afford support to vast herds of buffaloes, as well as
deer and other animals.

LAKES.

The most remarkable feature in North America is its lake-
system—the largest and most important in the world.   In
the north-west, at the foot of the Rocky Mountains, are the
Great Bear and Great Slave Lakes, which discharge their
waters through the Mackenzie River into the Arctic Ocean.
Next we have the Athabasca, Wollaston, and Deer Lakes.
In the very centre of the continent are the two important
lakes of Winnipeg and Winnipegoos,—the former 240 miles
in length by 55 in width, and the latter about half the size.
The large river of the Saskatchewan flows into Lake Winnipeg, and with it will, ere long, form an important means of
communication between the different parts of that vast district
lately opened up for colonization.   At its southern end the
Red River of the north flows into it, on the banks of which
a British settlement has long been established.   Several
streams, however, make their way into Hudson Bay.   Be-

tween it and Lake Superior is an elevated ridge of about
1500 feet in height; the streams on the west falling into
Lake Winnipeg, while those which flow towards the east reach
Lake Superior.

We now come to the site of the five largest fresh-water·
lakes in the world. Lake Superior extends, from west
to east, 335 miles, with an extreme breadth of 175. Its
waters flow through the St. Mary's River by a rapid descent
into Lake Huron, which is 240 miles long. This lake is
divided by the Manitoulin islands into two portions, and is
connected with Lake Michigan by a narrow channel without
rapids, so that the two lakes together may be considered to
form one sheet of water. On its southern extremity the
waters of Lake Huron flow through another narrow channel,
which expands during part of its course into Lake St. Clair;
and they then enter Lake Erie, which has a length of 265
miles, and a breadth of 80 miles. It is of much less depth than
the other lakes, and its surface is therefore easily broken up
into dangerous billows by strong winds. Passing onward
towards the north-east, the current enters the Niagara River,
about half-way down which it leaps along a rocky ledge ·of
100 feet in height, to a lower level, forming the celebrated
Falls of Niagara, and then passes on in a rapid course into
Lake Ontario. The fall between the two lakes is 333 feet.
Lake Ontario is 180 miles long and 65 miles wide. Out
of its north-eastern end falls the broad stream which here
generally takes the name of the St. Lawrence, and which
proceeds onward, now widening into lake-like expanses full of
islands, now compressed into a narrow channel, in a north-
easterly direction. The true St. Lawrence may indeed be
considered as traversing the whole system of the great lakes of

THE FALLS OF NIAGARA

North America, and thus being little less than a thousand miles
in direct length ; indeed, including its windings, it is fully two
thousand miles long.   To the north-west of it exist countless
numbers of small lakes united by a network of streams; while

SCENERY ON THE ST. LAWRENCE--LAKE OF THE THOUSAND ISLES.

numerous large rivers, such as the Ottawa, the St. Maurice,
and the Saguenay, flow into it, and assist to swell its current.
There are numerous other small lakes to the west of the
Rocky Mountains, a large number of which exist in the
Province of British Columbia, and are more or less connected
with the Fraser and Columbia Rivers.   Further to the south
are other lakes, many of them of volcanic origin, some
intensely salt, others formed of hot mud.   Among these is the

We must now proceed more particularly to examine the regions of which we have obtained the preceding cursory view, but, before we do so, we must glance at their human inhabitants.

ABORIGINAL INHABITANTS—THE RED MEN OF THE WILDS.

While the white men from Europe occupy the whole eastern coast, pressing rapidly and steadily westward, the Redskin aborigines maintain a precarious existence throughout the centre of the continent, from north to south, and are still found here and there on the western shores. On the northern ice-bound coast, the skin-clothed Esquimau wander in small bands from Behring Strait to Baffin Bay, but never venture far inland, being kept in check by their hereditary enemies, the Athabascas, the most northern of the red-skinned nations. The Esquimau, inhabiting the Arctic regions, may more properly be described in the volume devoted to that part of the globe.

INDIAN WIGWAMS.

Here and there, in openings in the primeval forest, either natural or artificial, on the banks of streams and lakes, several small conical structures may be seen, composed of long stakes, stuck in the ground in circular form, and fastened at the top.

The walls consist of large sheets of birch bark, layer above layer, fastened to the stakes. On the lee-side is left a small opening for ingress and egress, which can be closed by a sheet of bark, or the skin of a wild animal. At the apex, also, an aperture is allowed to remain for the escape of the smoke from the fire which burns within. Lines are secured to the

INDIAN WIGWAMS.

stakes within, on which various articles are suspended; while round the interior mats or skins are spread to serve as couches, the centre being left free for the fire. In front. forked stakes support horizontal poles, on which fish or skins are hung to dry; and against others, sheets of bark are placed on the weather-side, forming lean-tos, shelters to larger fires, used for more extensive culinary operations than can be carried on within the hut. On the shores are seen drawn up beautifully-formed canoes of birch bark of various sizes—some sufficient to carry eight or ten men; and others, in which only one or two people can sit.

APPEARANCE OF THE INDIANS.

Amid the huts may be seen human figures with dull copper or reddish-brown complexions, clothed in rudely-tanned skins of a yellowish or white hue, and ornamented with the teeth of animals and coloured grasses, or worsted and beads. Their figures are tall and slight. They have black, piercing eyes, slightly inclining downwards towards the nose, which is broad and large. They have thick, coarse lips, high and prominent cheek-bones, with somewhat narrow foreheads, and coarse, dark, glossy hair, without an approach to a curl; their heads sometimes adorned with feathered caps or other ornaments. Often their faces are besmeared with various coloured pigments in stripes or patches—one colour on one side of the face, the other being of a different hue. Their lower extremities are covered with leggings of leather, ornamented with fringes, and their feet clothed in mocassins of the same material as their leggings. The men stalk carelessly about, or repair their canoes or fishing gear and arms; while the women sit, crouching down to the ground, bending over their caldrons, shelling Indian corn, or engaged in some other domestic occupation; and the children, innocent of clothing, tumble about on the ground. In travelling, the Indian mother carries her child on her back. It is strapped to a board; and when a halting-place is reached, the cradle and the child are hung upon a tree, or on a pole inside the wigwam. Those who have communication with the whites may be seen clothed in blanket garments, which the men wear in the shape

INDIAN MODE OF CARRYING CHILDREN.

of coats; while the women swathe their bodies in a whole blanket, which covers them from their shoulders to their feet.

Though the men assume a grave and dignified air when a stranger approaches, they often indulge in practical jokes and laughter among themselves; and in seasons of prosperity, appear good-humoured and merry. The women, however, are doomed to lives of unremitting toil, from the time they become wives. They are compelled to carry the burdens, and to cultivate the ground, when any ground is cultivated, for the production of potatoes, maize, and tobacco. The men condescend merely to manufacture their arms and canoes, and to hunt; or they engage in what they consider the noblest of employments, waging war on their neighbours. The women, indeed, are often compelled to paddle the canoes, sometimes to go fishing, and to carry the portable property from place to place, or an overload of game when captured.

Intelligent as the Indian appears, it is evident that he has cultivated his perceptive powers to the neglect of his spiritual and moral qualities. His senses are remarkably acute. His memory is good; and when aroused, his imagination is vivid, though wild in the extreme. He is warmly attached to hereditary customs and manners. Naturally indolent and slothful, he detests labour, and looks upon it as a disgrace, though he will go through great fatigue when hunting or engaged in warfare.

WOOD INDIANS.

The northern tribes are known as Wood Indians, in contradistinction to the inhabitants of the open country, the Prairie Indians, who differ greatly from the former in their habits and customs. All the tribes of the Athabascas, as well as those to the south of them, known as the Algonquins, are Wood Indians. They are nearly always engaged in hunting the wild animals of the region they inhabit, for the sake of their furs, which they dispose of to the agents of the Hudson Bay Company and other traders, in exchange for blankets,

fire-arms, hatchets, and numerous other articles, as well as too
often for the pernicious fire-water, to obtain even small quantities of which they will frequently dispose of the skins
which it has cost them many weeks to obtain with much
hardship and danger.   These Wood Indians are peaceably
disposed, and can always escape the attacks of their enemies
of the prairies by retreating among their forest or lake

INDIANS SPEARING FISH.

fastnesses.   They obtain their game by various devices, sometimes using traps of ingenious construction, or shooting the
creatures with bows and arrows, and of later years with fire-
arms.   They spear the fish which abound in their waters, or
catch them with scoop and other nets.   Although their ordinary wigwams are of the shape already described, some are considerably larger, somewhat of a bee-hive form, covered thickly

with birch bark, and have a raised dais in the interior capable of holding a considerable number of people. The best known of these Forest Indians are the Chippeways, who range from the banks of Lake Huron almost to the Rocky Mountains, throughout the British territory.

THE PRAIRIE INDIANS.

To the south of the tribes already mentioned, are the large family of the Dakotahs, who number among them the Sioux, Assiniboines, and Blackfeet, and are the hereditary enemies of the Chippeways, especially of their nearer neighbours, the Crees and Ojibbeways. These Dakotahs occupy the open prairie country to the south of the Saskatchewan, and are the most northern of the Prairie Indians. In summer, they wear little or no clothing; and possessing numerous horses, hunt the buffaloes, or rather bisons, on horseback, armed with spears and bows and arrows. They are fiercer and more warlike than their northern neighbours, and have long set the whites at defiance. The buffalo supplies them with their chief support. The flesh of the animal dried in the sun, or pounded with its fat into pemmican, is their chief article of food; while its skin serves as a covering for their tents, a couch at night, or for clothing by day, and is manufactured into bags for carrying their provisions, and numerous other articles. Physically, they are superior to the Wood Indians. They are both hunters and warriors; and though they may occasionally exchange the buffalo robes—as the skins are called—for fire-arms, they seldom employ themselves as trappers, or attend to the cultivation of the ground.

The greater number of the tribes further to the south possess horses, and hunt the buffalo and deer. Some are even more savage than the Dakotahs, while others, again, have made slight progress towards civilization, and live in settled villages, while they rudely cultivate the ground, and possess herds of cattle.

Although the Indian languages differ greatly from each other, a great similarity in grammatical structure and form has been found to exist among them, denoting a common, though remote origin.   They differ, however, so greatly from any known language of the Old World, as to afford conclusive proof that their ancestors must have left its shores at an early period of the world's history.

The governments also differed.   In some tribes it approached an absolute monarchy, the will of the sachem or chief being the supreme law ; while in others it was almost entirely republican, the chief being elected for his personal qualities, though frequently the leadership was preserved in the female line of particular families.

When describing the customs of the Indians, we are compelled often to speak of the past, as the tribes, from being pressed together by the advancement of civilization, have become amalgamated, and many of their customs have passed away.   Most of the nations were divided into three or more clans or tribes, each distinguished by the name of an animal. Thus the Huron Indians were divided into three tribes—those of the Bear, the Wolf, and the Turtle.   The Chippeways, especially, were divided into a considerable number of tribes.

RELIGIOUS BELIEF.

Though their language differs so greatly, as do many of their customs, their religious notions exhibit great uniformity throughout the whole country.   They all possess a belief,

though it is vague and indistinct, in the existence of a Supreme, All-Powerful Being, and in the immortality of the soul,
which, they suppose, restored to its body, will enjoy the
future on those happy hunting-grounds which form the red
man's heaven. They also worship numerous inferior deities
or evil spirits, whom they endeavour to propitiate, under the
supposition that unless they do so they may work them evil
rather than good. They suppose that there is one god of
the sun, moon, and stars; that the ocean is ruled by another
god, and that storms are produced by the power of various
malign beings; yet that all are inferior to the Supreme Ruler
of the universe. We can trace in some of the tribes customs
and notions which have been derived from those of far distant nations. Thus, the tribes of Louisiana kept a sacred fire
constantly burning in their temples: the Natches, as did the
Mexicans, worshipped the sun, from whom their chiefs pretended to be descended. By some tribes human sacrifices
were offered up,—a custom which was practised by the
Pawnees and Indians of the Missouri even to a late period.
Several of the tribes buried their dead beneath their houses;
and it was an universal custom among all to inter them in a
sitting posture, clothed in their best garments, while their
weapons and household utensils, with a supply of food, were
placed in their graves, to be used when they might be restored to life. Several of their traditions evidently refer to
events recorded in Scripture history. The Algonquin tribes
still preserve one pointing to the upheaval of the earth from
the waters, and of a subsequent inundation. The Iroquois
have a tradition of a general deluge; while another tribe believe not only that a deluge took place, but that there was
an age of fire which destroyed all things, with the exception

of a man and woman, who were preserved in a cavern. Many similar traditions exist; while it is probable that those mentioned refer to the destruction of the Cities of the Plain by fire which came down from heaven, and to the confusion of tongues which fell upon the descendants of Noah in the plain of Shinar.

AMERICAN ANTIQUITIES.

We are apt to suppose that the wild inhabitants of the New World have ever existed in the same savage state as that in which they are found. Vast numbers, however, of remains, and buildings of great antiquity, have of late years been discovered, showing that at one time either their ancestors, or other tribes who have passed away, had made great progress in civilization. As the white man has advanced westward, and dug deep into the soil, whilst forming railway-cuttings, digging wells, and other works, numerous interesting remains have been discovered—a large number of fortified camps of vast extent, and even the foundations of cities, with their streets and squares, have been brought to light. Idols, pitchers of clay, ornaments of copper, circular medals, arrow-heads, and even mirrors of isinglass, in great numbers, have been found throughout the country. Some of the articles of pottery are skilfully wrought, and polished, glazed, and burned; inferior in no respects to those of Egypt and Babylon.

In Tennessee, an earthen pitcher, holding a gallon, was discovered on a rock twenty feet below the surface. It was surmounted by the figure of a female head covered with a conical cap. The features greatly resembled those of Asiatics, and the ears, extending as low as the chin, were of great size.

Near the Cumberland River an idol formed of clay was found about four feet below the surface of the earth. It is of curious construction, consisting of three hollow heads joined together at the back by an inverted bell-shaped hollow stem. This specimen also has strongly-marked Asiatic features; the red and yellow colour with which it is ornamented still retaining great brilliancy. Another idol, formed of clay and gypsum, was discovered near Nashville. It represented a human being without arms. The hair was plaited, and there was a band round the head with a flattened lump or cake upon the summit. Numerous medals, also, have been dug up, representing the sun, with its rays of light, together with utensils and ornaments of copper, sometimes plated with silver; and a solid silver cup, with its surface smooth and regular, and its interior finely gilt.

But besides these, and very many similar articles, throughout the whole country, and especially towards the west, immense numbers of fortresses of great size have been discovered, with walls of earth, some of them ten feet in height, and thirty in breadth. There is a vast fortress in Ohio, near the town of Newark. It is situated on an extensive plain, at the junction of two branches of the Musking-um. At the western extremity of the work stood a circular fort, containing twenty-two acres, on one side of which was an elevation thirty feet high, partly of earth and partly of stone. The circular fort was connected by walls of earth with an octagonal fort containing forty acres, the walls of which were ten feet high. At this end were eight openings or gateways about fifteen feet in width, each protected by a mound of earth on the inside. From thence four parallel walls of earth proceeded to the basin of the harbour, others

extending several miles into the country, and others on the
east joined to a square fort containing twenty acres, not
four miles distant. From this latter fort parallel walls extended to the harbour, and others to another circular fort one
mile and a half distant, containing twenty-six acres, and surrounded by an embankment from twenty-five to thirty feet
high. Further north and east the elevated ground was protected by intrenchments. Traces of other walls have been
found, apparently connecting these works with those thirty
miles distant. When we come to reflect that there were many
hundreds of similar forts, some of which were of equal size, and
others even of still greater magnitude, we cannot help believing
that an enormous population, considerably advanced in the arts
of civilization, must at one time have existed in the country,
over which for ages past the untutored savage has roamed in
almost a state of nature. And now these wild tribes are
rapidly disappearing before the advancement of a still greater
multitude, and a far more perfect civilization. Whether these
ancient races were the ancestors of the present Indians or not,
it is difficult to determine, as are the causes of their disappear-
ance. It is possible that, retreating southward, they established
the empires of Mexico and Peru, or, overcome by more savage
tribes, were ultimately exterminated.

CHAPTER II.

THE North American continent may be divided into four zones or parallel regions, which, from the difference in temperature which exists between them, present a great variety both in their fauna and flora.

THE FIRST ZONE.

Commencing on the east, where the Greenland Sea washes the coast of Labrador, and Hudson Strait leads to the intricate channels communicating with the Arctic Ocean, we have on the first-named coast a low and level region, which rises inland to a considerable elevation, and then once more sinks on the shores of Hudson Bay. West of that bay there is a wide extent of low country, intermixed with numerous lakes and marshes; and then along the Arctic shore is a wild, barren, treeless district, rising at length into the mountainous region of the Arctic highlands. Amid them numerous rapid streams find their way into the Arctic Ocean. Again they sink into the basin of the Mackenzie River, which separates them from the northern end of the Rocky Mountains. Hence westward to the Pacific is a broad highland region, rising into the lofty range of the Sea Alps.

THE SECOND ZONE—THE FERTILE BELT OF RUPERT'S LAND.

The next Zone we will consider as commencing at the Gulf of St. Lawrence. Westward extends an elevated region, rising in many places to a considerable height, and water-shed of the rivers which flow on the south St. Lawrence, and on the north into Hudson Bay. Proceeding up the St. Lawrence, we arrive at a great lake district, which embraces Lakes Ontario, Erie, Huron, Michigan, and Superior, to the extreme west. On the north-western shores of that lake we find an elevated district with several small lakes and streams flowing through valleys. This is the water-shed also of two systems. The streams to the east, flowing into Lake Superior, ultimately enter the St. Lawrence ; while those to the west make their way into Lake Winnipeg, the waters of which, after flowing through a variety of channels, fall into Hudson Bay. To the west of this water-shed range the first lake we meet with is known as the Lac des Milles Lacs. Two rivers flow from it, expanding here and there into small lakes, till another expanse of water is reached called Rainy Lake. This in the same way communicates by two streams with the still larger Lake of the Woods, the whole region on both sides being thickly wooded. From the Lake of the Woods flows the broad and rapid Winnipeg River, which finally falls into Lake Winnipeg. This large and long lake is con-nected with several others of smaller size,—Lake Winnipegoos and Manitoba Lake to the west of it. Into the southern end of Lake Winnipeg flows the Red River, which rises far away in the south in the United States, taking an almost direct

SCENE IN THE LAKE DISTRICT NORTH-WEST OF LAKE SUPERIOR.

northerly course. Towards the north, about twenty miles from the lake, is situated the well-known Selkirk settlement.

To the west of the Red River commences a broad belt of prairie land which extends here and there, rising into wooded heights and swelling hills, with several large rivers flowing through it, to the very base of the Rocky Mountains. As we advance westward we find it extending considerably to the north, where the large and wide river Saskatchewan, rising in the Rocky Mountains, flows eastward into Lake Winnipeg. Along the southern border of this region the Assiniboine River, also of considerable size, flows into the Red River at Fort Garry, in the Selkirk settlement. The prairie country indeed extends further than the Red River, up to the Lake of the Woods. The name of the FERTILE BELT has been properly given to it. Commencing at the Lake of the Woods, it stretches westward for 800 miles, and averages from 80 to upwards of 100 miles in width. The area of this extraordinary belt of rich soil and pasturage is about 40,000,000 of acres. Including the adjacent fertile districts, the area may be estimated at not less than 80,000 square miles, or considerably more fertile land than the whole of Canada is supposed to contain. It rises gradually towards the west, so that the traveller is surprised to find how speedily he has gained the passes which lead him over the Rocky Mountains into the territory of British Columbia on their western side—often indeed before he has realized the fact that he has crossed the boundary-line. The Fertile Belt is considerably more to the south than the British Islands, though, as the western hemisphere is subject to greater alternations of heat and cold than the eastern, there is a vast difference in temperature between the summer and winter. While in winter the whole

THE PRAIRIE OF THE FERTILE BELT.

region is covered thickly with snow, in summer the heat is so great that Indian corn and other cereals, as well as all fruits, ripen with great rapidity. The whole of this fertile region, which now forms part of the Canadian Dominion, is about to be opened to colonization; and through it will be carried the great high road which will connect the British provinces on the Pacific with those of the Atlantic.

ANIMAL LIFE ON THE FERTILE BELT.

Throughout this fine region range large herds of buffalo,— not extending their migrations, however, beyond its northern boundary. Here, too, are found two kind of small deer—the wapiti, and the prong-horned antelope. Hares—called rabbits, however—exist in great numbers. Porcupines are frequently found. The black bear occasionally comes out of the neighbouring forests, while a great variety of birds frequent the lakes and streams, whose waters also swarm with numerous fish. The white fish found in the lakes are much esteemed, and weigh from two or three to seven pounds. There are fine pike also. Sturgeon are caught in Lake Winnipeg and the Lower Saskatchewan of the weight of 160 pounds. Trout grow to a great size, and there are gold-eyes, suckers, and cat-fish. Unattractive as are the names of the two last, the fish themselves are excellent. Among the birds, Professor Hind mentions prairie-hens, plovers, various ducks, loons, and other aquatic birds, besides the partridge, quail, whip-poor-will, hairy woodpecker, Canadian jay, blue jay, Indian hen, and woodcock. In the mountain region are big-horns and mountain goats; the grizzly bear often descends from his rugged heights into the plains, and affords sport to the daring hunter. The musk-rat and beaver inhabit the borders of the lakes. The cariboo and moose frequent the Fertile Belt, though the musk-ox confines himself to the more northern regions. Wolves have been almost exterminated in the neighbourhood of the Red River settlement. The half-breeds and Indians possess peculiarly hardy and sagacious horses, which are trained for hunting the buffalo. Their dogs are large and powerful, and four of them will draw a sleigh with one man over the snow at the rate of six miles an hour. Herds of cattle, as well as horses and hogs, are left out during the whole winter, it being necessary only—should a thaw come on, succeeded by a frost—to supply them with food ; otherwise, unable to break through the coating of ice thus formed, they are liable to starve.

The farmers of the Red River settlement grow wheat, barley, oats, flax, hemp, hops, turnips, and even tobacco, though Indian corn grows best, and can always be relied on. Wheat, however, is the staple crop of Red River. It is a splendid country for sheep pasturage, and did easier means of transporting the wool exist, or could it be made into cloth or blankets in the settlement, no doubt great attention would be given to the rearing of sheep.

THE THIRD ZONE—THE DISMAL SWAMP IN THE UNITED STATES.

Returning again to the east coast, about the latitude of Chesapeake Bay and Cape Hatteras, we find a low level region known as the Atlantic plain, running parallel to the coast, on which the long-leaved or peach-pines flourish. This region is generally called the Pine Barrens. Wild vines encircle the trees, and among them are seen the white berries of the mistletoe. In winter these Pine Barrens retain much of their verdure, and constitute one of the marked features of the

country.   Amid them are numerous swamps or morasses.   One
of great size, extending to not less than forty miles from north
to south, and twenty-five in its greatest width, is called the
GREAT DISMAL SWAMP.   The soil, black as in a peat-bog, is
covered with all kinds of aquatic trees and shrubs; yet,
strange to say, instead of being lower than the level of the
surrounding country, it is in the centre higher than towards
its margin; indeed, from three sides of the swamp the waters
actually flow into different rivers at a considerable rate.
Probably the centre of the morass is not less than twelve feet
above the flat country around it.   Here and there some
ridges of dry land appear, like low islands, above the general
surface.   On the west, however, the ground is higher, and
streams flow into the swamp, but they are free from sediment,
and consequently bring down no liquid mire to add to its
substance.   The soil is formed completely of vegetable matter,
without any admixture of earthy particles.   In many even
of the softest parts juniper-trees stand firmly fixed by their
long tap roots, affording a dark shade, beneath which numerous ferns, reeds, and shrubs, together with a thick carpet of
mosses, flourish, protected from the rays of the sun.   Here and
there also large cedars and other deciduous trees have grown up.
The black soil formed beneath, increased by the rotting vegetation, is quite unlike the peat of Europe, as the plants become
so decayed as to leave no traces of organization.   Frequently
the trees are overthrown, and numbers are found lying beneath
the surface of the soil, where, covered with water, they never
decompose.   So completely preserved are they, that they are
frequently sawn up into planks.   In one part of the Dismal
Swamp there is a lake seven miles in length, and more than
five wide, with a forest growing on its banks.   The water is

transparent, though tinged with a pale brown colour, and contains numerous fish. The region is inhabited by a number of bears, who climb the trees in search of acorns and gum-berries, breaking off the boughs of the oaks in order to obtain the acorns; these bears also kill hogs, and even cows. Occasionally a solitary wolf is seen prowling over the morass, and wild cats also clamber amid its woods. Even in summer, the air, instead of being hot and pestiferous, is especially cool, the evaporation continually going on in the wet spongy soil generating an atmosphere resembling that of a region consider-ably elevated above the level of the ocean. Canals have been cut through this swamp. They are shaded by tall trees, their branches almost joining across, and throwing a dark shade on the water, which itself looks almost black, and adds to the gloom of the region. Emerging from one of these avenues into the bright sunlit lake, the aspect of the scenery is like that of some beautiful fairyland.

FOSSIL FOOTMARKS OF BIRDS.

A considerable way to the north of this region, on the banks of the Connecticut River, are beds of red sandstone, on the different layers of which are found the footmarks of long extinct birds. The beds in some parts are twenty-five feet in thickness, composed of layer upon layer; and on each of these layers, when horizontally split, are found imprinted these remarkable footmarks. This result could only have been produced by the subsidence of the ground, fresh deposi-tions of sand having taken place on the layers, on which the birds walked after the subsidence. They must have been of various sizes,—some no larger than a small sand-piper, while others, judging from their footprints, which measure no less than nineteen inches, must have been twice the size of the modern African ostrich.   The distances between the smaller measure only about three inches, but in the case of the largest, called the ORNITHICHNITES GIGAS, they are from four to six feet apart.   In some places where the birds have congregated together none of the steps can be distinctly traced, but at a short distance from this area the tracks become more and

FOSSIL FOOTMARKS OF BIRDS.

more distinct.   Upwards of two thousand such footprints have been observed, made probably by nearly thirty distinct species of birds, all indented on the upper surface of the strata, and only exhibiting casts in relief on the under side of the beds which rested on such indented surfaces.   In other places the marks of rain and hail which fell countless ages ago are clearly visible.   Sir Charles Lyell perceived similar footprints in the red mud in the Bay of Fundy, which had just been formed by sand-pipers; and on examining an inferior layer of mud, formed several tides before, and covered up by fresh sand, he discovered casts of impressions similar to those made on the last-formed layer of mud. Near the footsteps he observed the mark of a single toe, occurring occasionally, and quite isolated from the rest. It was suggested to him that these marks were formed by waders, which, as they fly near the ground, often let one leg hang down, so that the longest toe touches the surface of the mud occasionally, leaving a single mark of this kind. He brought away some slabs of the recently formed mud, in order that naturalists who were sceptical as to the real origin of the ancient fossil ornithichnites might compare the fossil products lately formed with those referable to the feathered bipeds which preceded the era of the ichthyosaurus and iguanodon.

THE BIG-BONE LICK.

We will now cross the Alleghanies westward, where we shall find a thickly-wooded country. As we proceed onwards, entering Kentucky, we reach a spot of great geological interest, called the BIG-BONE LICK. These licks exist in various parts of the country. They are marshy swamps in which saline springs break out, and are frequented by buffalo, deer, and other wild animals, for the sake of the salt with which in the summer they are incrusted, and which in winter is dissolved in the mud. Wild beasts, as well as cattle, greedily devour this incrustation, and will burrow into the clay impregnated with salt in order to lick the mud. In the Big-bone Lick of Kentucky the bones of a vast number of mastodons and other extinct quadrupeds have been dug up.

This celebrated bog is situated in a nearly level plain, bounded by gentle slopes, which lead up to wide-extended table-lands. In the spots where the salt springs rise, the bog is so soft that a man may force a pile into it many yards perpendicularly. Some of these quaking bogs are even now more than fifteen acres in extent, but were formerly much larger, before the surrounding forest was partially cleared away. Even at the present day cows, horses, and other quadrupeds are occasionally lost here, as they venture on to the treacherous ground. It may be easily understood, therefore, how the vast mastodons, elephants, and other huge animals lost their lives. In their eagerness to drink the saline waters, or lick the salt, those in front, hurrying forward, would have been pressed upon by those behind, and thus, before they were aware of their danger, sank helplessly into the quagmire. It is supposed that the bones of not less than one hundred mastodons and twenty elephants have been dug up out of the bog, besides which the bones of a stag, extinct horse, megalonyx, and bison, have been obtained. Undoubtedly, therefore, this plain has remained unchanged in all its principal features since the period when these vast extinct quadrupeds inhabited the banks of the Ohio and its tributaries. Here and there the Big-bone Lick is covered with mud, washed over it by some unusual rising of the Ohio River, which is known to swell sixty feet above its summer level.

Passing on through wide-spreading prairies, we cross the mighty stream of the Mississippi to a slightly elevated district of broad savannahs, till we reach a treeless region bordering the very foot of the Rocky Mountains. Through this region numerous rivers pass on their way to the Mississippi. Leaving at length the great western plain, we begin to mount the slopes of the Rocky Mountains, when we may gaze upwards at the lofty snow-covered peaks above our heads. Hence, crossing the mighty range in spite of grizzly bears and wilder Indians, we descend towards the bank of the Rio Colorado, which falls into the Gulf of California, and thence over a mountainous region, some of whose heights, as Mount Dana, reach an elevation of 13,000 feet, and Mount Whitney, 15,000 feet.

THE FOURTH ZONE.

The southernmost of the four zones begins on the coast of Florida, passes for hundreds of miles over a low or gently sloping country toward the great western plains which border the Rocky Mountains into Texas ; its southern boundary being the Gulf of Mexico. Through this region flow numerous rivers, the queen of which is the Mississippi. The western portion is often wild and barren in the extreme, inhabited only by bands of wild and savage Indians. The Rocky Mountains being passed, there is a lofty table-land, and then rise the Sierras de los Nimbres and Madre ; beyond which, bordering the Gulf of California, is the wild, grandly pictur-esque province of Sonora, with its gigantic trees and stalactite caves.

CHAPTER III.

TO obtain, however, a still more correct notion of the appearance of the continent, we must take another glance over it. We shall discover, to the north, and throughout the eastern portion where civilized man has not been at work clearing away the trees, a densely-wooded region. Proceeding westward, as the valley of the Mississippi is approached the underwood disappears, and oak openings predominate. These OAK OPENINGS, as they are called, are groves of oak and other forest trees which are not connected, but are scattered over the surface at a considerable distance from one another, without any low shrub or underbrush between them.

THE PRAIRIES.

Thus, gradually, we are entering the prairie country, which extends as far west as the Grand Coteau of the Missouri. This prairie region is covered with a rich growth of grass; the soil is extremely fertile, and capable of producing a variety of cereals. Over the greater portion of the prairie country, indeed, forests of aspens would grow. did not annual fires in most parts arrest their progress. Here and there numbers have sprung up. The true prairie region in the United States extends over the eastern part of Ohio, Indiana, the southern portion of Michigan, the southern part of Wisconsin, nearly the whole of the states of Illinois and Iowa, and the northern portion of Missouri, gradually passing—in the territories of Kansas and Nebraska—into that arid and desert region known as THE PLAINS, which lie at the base of the Rocky Mountains.

The Grand Coteau de Missouri forms a natural boundary to these arid plains. This vast table-land rises to the height of from 400 to 800 feet above the Missouri. Vegetation is very scanty; the Indian turnip, however, is common, as is also a species of cactus. No tree or shrub is seen; and only in the bottoms or in marshes is a rank herbage found. Across these desert regions the trails of the emigrant bands passing to the Far West have often been marked : first, in the east, by fur- niture and goods abandoned ; further west, by the waggons and carts of the ill-starred travellers ; then by the bones or oxen and horses bleaching on the plain ; and, finally, by the graves, and sometimes the unburied bodies, of the emigrants themselves, the survivors having been compelled to push onwards with the remnant of their cattle to a more fertile region, where provender and water could be procured to restore their well-nigh exhausted strength. Oftentimes they have been attacked by bands of mounted Indians, whose war-whoop has startled them from their slumbers at night; and they have been compelled to fight their way onwards, day after day assailed by their savage and persevering foes.

Civilized man is, however, triumphant at last, and the steam-engine, on its iron path, now traverses that wild region from east to west at rapid speed ; and the red men, who

claim to be lords of the soil, have been driven back into the
more remote wilderness, or compelled to succumb to the

INDIANS ATTACKING AN EMIGRANT TRAIN.

superior power of the invader, in many instances being utterly
exterminated.   Still, north and south of that iron line the
country resembles a desert; and the wild Indian roams as of
yore, like the Arab of the East—his hand against every man,
and every man's hand against him.

Among the dangers to which the traveller across the prairie is exposed, the most fearful is that of fire. The sky is bright overhead; the tall grass, which has already assumed a yellow tinge from the heat of summer, waves round him, affording abundant pasture to his steed. Suddenly his guides rise in their stirrups and look anxiously towards the horizon. He sees, perhaps, a white column of smoke rising in the clear air. It is so far off that it seems it can but little concern them. The guides, however, think differently, and after a moment's consultation point eagerly in the direction of some broad river, whose waters flow towards the Mississippi. "Onward! onward!" is the cry. They put spurs to their horses' flanks, and gallop for their lives. Every instant the column of smoke increases in width, till it extends directly across the horizon. It grows denser and denser. Now above the tall grass flashes of bright light can be seen. The traveller almost fancies he can hear the crackling of the flames as they seize all combustible substances in their course. Now they surround a grove of aspens, and the fierce fire blazes up more brightly than ever towards the sky, over which hangs a dark canopy of smoke. Suddenly a distant tramp of feet is heard. The very ground trembles. A dark mass approaches—a phalanx of horns and streaming manes. It is a herd of buffaloes, turned by the fire purposely ignited by the Indians. The guides urge the travellers to increase their speed; for if overtaken by the maddened animals, they will be struck down and trampled to death. Happily they escape the surging herd which comes sweeping onward— thousands of dark forms pressed together, utterly regardless of the human beings who have so narrowly escaped them. The travellers gallop on till their eyes are gladdened by the

sight of the flowing waters of a river. They rush down the
bank. Perchance the stream is too rapid or too deep to be
forded. At the water's edge they at length dismount, when
the Indians, drawing forth flint and steel, set fire to the grass
on the bank. The smoke well-nigh stifles them, but the flames
pass on, clearing an open space; and now, crouching down
to the water's edge, they see the fearful conflagration rapidly
approaching. The fire they have created meets the flames
which have been raging far and wide across the region.
And now the wind carries the smoke in dense masses over
their heads; but their lives are saved—and at length. they
may venture to ride along the banks, over the still smoulder-
ing embers, till a ford is reached, and they may cross the
river to where the grass still flourishes in rich luxuriance.

While, on one side of the stream, charred trees are seen
rising out of the blackened ground, on the other all is green
and smiling. These fearful prairie fires, by which thousands
of acres of grass and numberless forests have been destroyed,
are almost always caused by the thoughtless Indians, either
for the sake of turning the herds of buffaloes towards
the direction they desire them to take, or else for signals
made as a sign to distant allies. Sometimes travellers have
carelessly left a camp-fire still burning, when the wind has
carried the blazing embers to some portion of the surrounding dry herbage, and a fearful conflagration has been the
result.

Mr. Paul Kane, the Canadian artist and traveller, mentions
one which he witnessed from Fort Edmonton. The wind
was blowing a perfect hurricane when the conflagration was
seen sweeping over the prairie, across which they had passed
but a few hours before. The night was intensely dark, add-

PRAIRIE ON FIRE.

ing effect to the brilliancy of the flames, and making the
scene look truly terrific.   So fiercely did the flames rage,
that at one time it was feared the fire would cross the river
to the side on which the fort is situated, in which case it and
all within must have been destroyed.   The inmates also had
had many apprehensions for the safety of one of their party,
from whom, with his Indians, Mr. Kane had parted some time
before, and who had not yet arrived.   For three days they
were uncertain of his fate, when at length their anxiety was
relieved by his appearance.   He had noticed the fire at a
long distance, and had immediately started for the nearest
bend in the river.   This, by great exertion, he had reached
in time to escape the flames, and had succeeded in crossing.

THE BARREN PLAINS IN THE FAR WEST.

On the prairies of the east the eye ranges over a wide
expanse of waving grass, everywhere like the sea.   As, crossing the plains, we proceed west towards the vast range of
the Rocky Mountains, the country gives evidence of the
violent and irregular disturbances to which it has been subjected.   Wild rocky ridges crop out from the sterile plains of
sand ; and for hundreds of miles around the country is desert,
dry, and barren.   Even the vegetation, such as it is, is of
the same unattractive character.   The ground here and there
is covered with patches of the gray gramma grass, growing
in little cork-screw curls ; and there is a small furzy plant,
the under sides of the leaves of which are covered with a
white down, while occasionally small orange-coloured flowers
are seen struggling into existence.

There are insects, however.   Ants swarm in all directions,
building cones a foot in height.   Grasshoppers in myriads,

with red wings and legs, fly through the air—the only bright objects in the landscape. Sometimes the reddish-brown cricket is seen. Even the Platte River, which flows through this region, partakes of its nature. It seems to consist of a saturated solution of sand : when a handful is taken up, a gray mud of silex remains in the palm. Dry as this gramma grass appears, it possesses nutritive qualities, as the animals which feed on it abundantly prove.

Storms break over these plains with tremendous fury : the thunder roars, the lightning which flashes from the clouds illumines earth and sky with a brightness surpassing the cloudless noon. Then again utter darkness covers the earth, when suddenly a column of light appears, like the trunk of some tall pine, as the electric fluid passes from the upper to the lower regions of the world. The next instant its blazing summit breaks into splinters on every side. Occasionally fearful hail-storms sweep over the plains ; and at other times the air from the south comes heated, as from a furnace, drying up all moisture from the skin, and parching the traveller's tongue with thirst.

Here and there are scattered pools of water containing large quantities of salts, soda, and potash, from drinking which numbers of cattle perish. The track of emigrants is strewn for many miles with bleaching heads, whole skeletons, and putrefying carcasses—the result of the malady thus produced, in addition to heat and overdriving. Even the traveller suffers greatly, feeling as if he had swallowed a quantity of raw soda.

Yet often in this generally desert region, where the rivers wind their way through the plain, or wide pools of pure water mirror the blue sky, scenes of great beauty are presented

Nothing can surpass the rosy hues which tinge the heavens at sunrise. Here game of all sorts is found. The lakes swarm with mallards, ducks, and a variety of teal. Herds of antelopes cross the plain in all directions, and vast herds of buffalo darken the horizon as they sweep by in their migrations.

THE ROCKY MOUNTAINS.

At length a blue range, which might be taken for a rising vapour, appears in the western horizon. It is the first sight the traveller obtains of the long-looked-for Rocky Mountains; yet he has many a weary league to pass before he is among them, and dangers not a few before he can descend their western slopes. At length he finds himself amid masses of dark brown rocks, not a patch of green appearing; mountain heights rising westward, one beyond the other; and far away, where he might suppose the plains were again to be found, still there rises before him a region of everlasting snow. For many days he may go on, now climbing, now descending, now flanking piles of rocks, and yet not till fully six days are passed is he able to say that he has crossed that mountain-range. Indeed, the term "range" scarcely describes the system of the Rocky Mountains. It is, in fact, a chain, composed of numerous links, with vast plains rising amid them.

PARKS.

These ranges in several places thin out, as it were, leaving a large tract of level country completely embosomed in snowy ridges in the very heart of the system. These plains are known as "parks." They are found throughout the range. Several of them are of vast extent,—the four principal ones forming the series called, in their order,

SCENERY AMONG THE ROCKY MOUNTAINS.

"North," "Middle," "South," and "St. Louis" Parks.
Portions of them, thoroughly irrigated, remain beautifully
green throughout the year, and herbage over the whole region
is abundant.   Sheltered from the blasts to which the lower
plains are exposed, these parks enjoy an equable climate;
and old hunters, who have camped in them for many seasons,
describe life there as an earthly paradise.   They abound in
animals of all sorts.   Elk, deer, and antelope feed on their
rich grasses.   Hither also the puma follows its prey, and there
are several other creatures of the feline tribe.   Bears, wolves,
and foxes likewise range across them.   In some of them herds
of buffalo pass their lives; for, unlike their brethren of the
plain, they are not migratory.   It is doubtful whether or not
they are of the same species, but they are said to be larger
and fiercer.

The appropriate designation of the Rocky Mountain-system is
that of a chain.   On crossing one of its basins or plateaux, the
traveller finds himself within a link such as has just been described.   A break in one of these links is called a " pass," or
" cañon."   As he passes through this break he enters another
link, belonging to another parallel either of a higher or lower
series.   In some of the minor plateaux between the snowy
ridges no vegetation appears.   Granite and sandstone rocks
outcrop even in the general sandy level, rising bare and perpendicularly from 50 to 300 feet; as a late traveller describes it, " looking like a mere clean skeleton of the world."
Nothing is visible but pure rock on every side.   Vast stones
lie heaped up into pyramids, as if they had been rent from
the sky.   Cubical masses, each covering an acre of surface,
and reaching to a perpendicular height of thirty or forty feet,
suggest the buttresses of some gigantic palace, whose super-

structure has crumbled away with the race of its Titanic builders. It is these regions especially which have given the mighty range the appropriate name of the *Rocky Mountains.*

• THE SAGE COCK.

In some spots, the limitless wastes are covered by a scrubby plant known as mountain sage. It rises from a tough gnarled root in a number of spiral shoots, which finally form a single trunk, varying in circumference from six inches to two feet. The leaves are gray, with a strong offensive smell resembling true sage. In other places there appear mixed with it the equally scrubby but somewhat greener grease-wood—the two resinous shrubs affording the only fuel on which the emigrant can rely while following the Rocky Mountain trail.

These sage regions are the habitation of a magnificent bird —the Sage Cock. He may well be called the King of the grouse tribe. When stalking erect through the sage, he looks as large as a good-sized wild turkey—his average length being, indeed, about thirty-two inches, and that of the hen two feet. They differ somewhat, according to the season of the year. The prevailing colour is that of a yellowish-brown or warm gray, mottled with darker brown, shading from cinnamon to jet black. The dark spots are laid on in a longitudinal series of crescents. The under parts are a light gray, sometimes almost pure white, barred with streaks of brown, or pied with black patches. In the elegance of his figure and fineness of his outlines he vies with the golden pheasant. His tail differs from that of the grouse family in general by coming to a point instead of opening like a fan. On each side of his neck he has a bare orange-coloured spot, and near it a downy epaulet. His call is a rapid " Cut, cut, cut !"

followed by a hollow blowing sound. He has the partridge's habit of drumming with his wings, while the hen-bird knows the trick of misleading the enemy from her young brood. He seldom rises from the ground, his occasional flights being low, short, and laboured. He runs with great speed, and in his favourite habitat dodges and skulks with rapidity, favoured by the resemblance of his colour to the natural tints of the scrub. Though sometimes called the Cock of the Plains, he never descends into the plains, being always found on the higher mountain regions.

When the snow begins to melt, the sage hen builds in the bush a nest of sticks and reeds artistically matted together, and lays from a dozen to twenty eggs, rather larger than those of the domestic fowl, of a tawny colour, irregularly marked with chocolate blotches on the larger end. When a brood is strong enough to travel, the parents lead their young into general society. They are excessively tame, or bold. Often they may be seen strutting between the gnarled trunk and ashen masses of foliage peculiar to the sage scrub, and paying no more attention to the traveller than would a barn-yard drove of turkeys ; the cocks now and then stopping to play the dandy before their more Quakerly little hens, in-flating the little yellow pouches of skin on either side of their necks, till they globe out like the pouches of a pigeon.

WINTER SCENE AMONG THE ROCKY MOUNTAINS.

Descending the precipitous slopes of the Rocky Mountains on the west, we enter on a vast plain no less than 2000 miles in length, though comparatively narrow—the great basin of California and Oregon. Its greatest width, from the Sierra Nevada to the Rocky Mountains, is nearly 600 miles, but is generally much less.   The largest lake found on it is 4200 feet above the level of the sea, and is connected with the Salt Lake of Utah.   The mean elevation of the plain is about 6000 feet above the sea.   A mountain-chain runs across it, and through it flows the large Colorado River, amidst gorges of the most picturesque magnificence.

If the scenes we have described are stern and forbidding in summer, how much more so are they in winter, when icy blasts blow through the cañons, and masses of snow cover the ground.   From one of the outer spurs on the east, let us take a glance over the region.   Behind us rises the chain of the Rocky Mountains, the whole intermediate country, as well as the mountains themselves, except where the precipitous rocks forbid it, being covered thickly with snow.   Rugged peaks and ridges, snow-clad and covered with pines, and deep gorges filled with broken rocks, everywhere meet the eye. To the east, the mountains gradually smooth away into high spurs and broken ground, till they join the wide-spreading plains, generally stretching far as the eye can reach, and hundreds of miles beyond—a sea of barrenness, vast and dismal. A hurricane blows clouds of white snowy dust across the desert, resembling the smoke of bonfires, roaring and raving through the pines on the mountain-top, filling the air with snow and broken branches, and piling it in huge drifts against the trees.

The perfect solitude of this vast wilderness is appalling. From our lofty post on the mountain-top, we obtain a view over the rugged and chaotic masses of the stupendous chain, and the vast deserts which stretch away far from its eastern base ; while on all sides are broken ridges and chasms and ravines, with masses of piled-up rocks and uprooted trees, with clouds of drifting snow flying through the air, and the hurricane's roar battling through the forest at our feet adding to the wildness of the scene, which is unrelieved by the slightest vestige of animal or human life.

THE HORNED FROG.

We must now pass in review some of the numerous animals which inhabit these regions. In some of the mountain pla- teaux, among the cactuses and sand-heaps, we find that singu- larly-made animal known vulgarly as the Texan toad or horned frog—a name which in no way properly belongs to him, as he is more nearly related to the lizards and sala- manders. He lives as contentedly on the hot baked prairies of Texas, as amongst their snow-surrounded heights; though, from his appearance, we should expect to see him basking under a semi-tropical sun, rather than in this region. Yet here he lives, and must often have to spend much of his time under the snow. These toads, as the creatures are called, have brown backs, white bellies, small twinkling black eyes, set in almond-shaped slits, enclosed by two dark marks of the same shape. This has the effect of enlarging the eye, and giving it a soft look like that of the antelope. The two retro-curved horns, which rise out of bony sockets above the eyes, add still more to this odd resemblance.

The skin of the back and the long stiff tail, instead of being warted like the true toad's upper surface, is set with thorny excrescences. That of the lower surface is a dry tough tissue, almost horny. Whether this armour is given him to defend himself from the rattlesnake, it is difficult to say. The crea- ture itself is of a peaceable disposition; and so unwilling is he to fight, that he will allow himself to be taken in the hand,

and if placed on it directly after capture, he will not attempt to get away. It is very easy to catch him in the first place, for his movements over the loose sand of his haunts are scarcely faster than those of a land tortoise.

The trappers and other scattered inhabitants of this region describe a fish with hands as frequenting the brooks and pools. Though there are, no doubt, some curious fish, it is questionable how far these creatures possess the members ascribed to them.

FUR-TRAPPERS OF THE FAR WEST.

The fur-trapper of America is the chief pioneer of the Far West. His life spent in the remote wilderness, with no other companion than Nature herself, his character assumes a mixture of simplicity and ferocity. He knows no wants beyond the means of procuring sufficient food and clothing. All the instincts of primitive man are constantly kept alive. Exposed to dangers of all sorts, he becomes callous to them, and is as ready to destroy human as well as animal life as he is to expose his own. He cares nothing for laws, human or divine. Strong, active, hardy, and daring, he depends on his instinct for the support of life.

The independent trapper possesses traps and animals of his own, ranges wherever he lists through the country, and disposes of his peltries to the highest bidder. There are others employed by the fur companies, who supply them with traps and animals, and pay a certain price for the furs they bring.

The independent trapper equips himself with a horse and two or three mules—the one for the saddle, the others for his packs—and a certain number of traps, which he carries in a leather bag, with ammunition, a few pounds of tobacco, and dressed deer-skins for his mocassins and repairing his gar-

ments.   His costume is a hunting-shirt of dressed buckskin,
ornamented with long fringes ; pantaloons of the same material, decorated with porcupine quills and long fringes down
the outside of the leg.   He has mocassins on his feet, and a
flexible felt hat on his head.   Over his left shoulder and
under his right arm hang his powder-horn and bullet-pouch,
with flint, steel, and other articles, in a bag.   A belt round
the waist secures a large knife in a sheath of buffalo hide to
a steel chain, as also a case of buckskin, containing a whet-
stone.   In his belt is also stuck a tomahawk, a pipe-holder
hangs round his neck, and a long heavy rifle is slung over his
shoulder.

Arrived on the hunting-ground, as soon as the ice has
broken up he follows the creeks and streams, keeping a look-
out for the signs of beavers.   As soon as he discovers one, he
sets his trap, secured to a chain fastened to a stake or tree,
baiting it with the tempting castoreum.   He is ever on the
watch for the neighbourhood of Indians, who try to outwit
him, though generally in vain, to steal his traps and beavers.
His eye surveys the surrounding country, and instantly de-
tects any sign of his foes.   A leaf turned down, the slightly
pressed grass, the uneasiness of the wild animals, the flight of
birds, all tell him that other human beings are in the neighbourhood.   Sometimes, after he has set his traps and is returning to his camp, the wily Indian who has been watching
follows, and a home-drawn arrow, shot within a few feet,
never fails to bring the hapless victim to the ground.   For
one white scalp, however, that dangles in the smoke of an
Indian's lodge, a dozen black ones surround the camp-fires of
the trappers' rendezvous.   Here, after the hunt, from all
quarters the hardy trappers bring in their packs of beaver

FUR-TRAPPER.

to meet the purchasers, sometimes to the value of a thousand dollars each. The traders sell their goods at enormous profits; and the thoughtless trapper, indulging in the fire-water from which he has long abstained, is too often induced to gamble away the gold for which he has risked life and gone through so many hardships. When all is gone, he gets credit for another equipment, and sets off alone, often to return and repeat the same process, although the profits of one or two successful hunts would enable him to stock a farm and live among civilized men.

WONDERS OF NATURE.——MAMMOTH CAVE OF· KENTUCKY.

There are many other wonders of Nature in different parts of North America well worthy of more notice than we can

THE DEAD SEA, MAMMOTH CAVE

give them. The most remarkable, perhaps, is the Mammoth Cave of Kentucky. The entrance to it is situated near Green River, midway between Louisville and Nashville. A lonely road leads to the entrance, from which, as we approach it in summer, we find a peculiarly chilly air issue forth. The sombre gloom of the entrance does not prepare us for the enormous hall within; long avenues leading into vast chambers, the smaller, thirty feet in height, at least, with an area of half an acre, and, as we get lower and lower, increasing in height. Upwards of eighteen miles of the cavern have been explored, and it may possibly be of still greater extent. To give an idea of the height of one of the chambers, we may add that the rocks from above have fallen, and a hill has been formed one hundred feet in eleva-

EYELESS FISH—
FRONT VIEW.

EYELESS FISH—
SIDE VIEW.

tion. Many of the halls are ornamented with the most magnificent stalactites. One of them is appropriately called

Martha's Vineyard, in consequence of having its tops and sides covered with stalactites which resemble bunches of grapes.

RIVER STYX, MAMMOTH CAVE, KENTUCKY.

Several streams pass through the cavern, down the sides of which rush numerous cataracts. Some of these streams, which are of considerable depth and width, are inhabited by shoals of eyeless fish, the organs of sight being superfluous in a region doomed to eternal night. The atmosphere of this huge cave is peculiarly dry, and is supposed to be extremely serviceable to persons afflicted with pulmonary complaints.

To visit any considerable portion of the cavern would occupy us at least a couple of days. It is calculated there are no less than two hundred and twenty-six avenues, forty-seven domes, numerous rivers, eight cataracts, and twenty-three pits,—many of which are grand in the extreme. Some of the rivers are navigated by boats, and, as may be supposed, they have obtained appropriate names. Here we find the Dead Sea and the River Styx. One of the streams disappears beneath the ground, and then rises again in another portion of the cavern. But after all, as naturalists, the little eyeless fish should chiefly claim our attention.

OIL SPRINGS.

As coal was stored up for the use of man, formed in ages past from the giant vegetation which then covered the face of the earth, so the Creator has caused to be deposited in sub-terranean caverns large quantities of valuable oil, which not only serves man for light, but is useful to him for many other purposes.

Whether that oil was produced from animal or vegetable substances, appears, even now, a matter of dispute. Some naturalists suppose that vast numbers of oil-giving creatures had been assembled in the districts in which these oil wells are now found, and the oil was pressed out of them by a superincumbent weight of rock. Others assert that the same result might be produced from a vast mass of oil-giving vegetation having been crushed by a similar process. Be that as it may, in several parts of the States, as well as in Canada, enormous pits exist full of this curious oil. It is obtained by boring in the ground in those spots where the oil is likely to

A ROCK-OIL SPRING.

be found. Often, however, the speculator, after spending time and capital in the experiment, finds that no oil appears at his call.

In some spots, where it was first discovered, after the

boring was completed, some hundreds of tons flowed up so
rapidly, that it was difficult to find casks sufficient to preserve
the produce. The whole region round is impregnated with
the odour of the oil. Long teams of waggons come laden with
casks of oil on the roads approaching the wells. Sheds for
repairing the casks, and storing the oil, are ranged around.
Every one gives indubitable signs by their appearance of their
occupation, while rock-oil, as it is called, is the only subject
of conversation in the neighbourhood.

MAMMOTH TREES AND CAVERNS OF CALAVERAS.

Gigantic as are the trees found in many of the eastern
forests of America, they are far surpassed by groves of pines
discovered a few years back in the southern parts of California.
They are found in small groves together—in some places only
three or four of the more gigantic in size; in others, as many
as thirty or forty, one vying with the other in height and
girth. In one grove, upwards of one hundred trees were found,
of great size, twenty of which were about seventy-five feet
in circumference. One of these trees, of greater size than its
companions, was sacrilegiously cut down. Its height was
302 feet, and its circumference, at the ground, 96 feet.
As it was impossible to *cut* it down, it was bored off with
pump-augers. This work employed five men for twenty-two
days. Even after the stem was fairly severed from the
stump, the uprightness of the tree and breadth of its base
sustained it in its position, and two days were employed in
inserting wedges and driving them in; but at length the
noble monarch of the forest was forced to tremble, and then
to fall, after braving the battle and the breeze for nearly
three thousand winters.

Many of the trees have received appropriate names. One has fallen, and has been hollowed out by fire. Through it a person can ride on horseback for sixty feet. Its estimated height, when standing, was 330 feet, and its circumference, 97 feet. Another of these giants is known as Hercules. It is 320 feet high, and 95 feet in circumference. Perhaps the most beautiful group is that of three trees known as the Three Graces. Each of them measures 92 feet in circumference at the base; and in height they are nearly equal, measuring 295 feet. Time was when, perhaps, the whole forest consisted of trees of the same size; but many have been destroyed by fire, and the time may come when none of those now standing will remain. The name of Wellingtonia has been given to the species.

MAMMOTH TREE OF CALIFORNIA.

In the same region are numerous magnificent stalactite caverns, which equal in beauty, if not in size, those of the Mammoth Cave of Kentucky

There are several waterfalls, unsurpassed for picturesque beauty.

YOSEMITE FALL, CALIFORNIA.

Had we time, we might pay a visit also to the gold-mines of California, and observe the way they are worked; but we should be prevented from giving that attention to the animal creation which is our present object.

CHAPTER IV.

GENERAL SURVEY OF THE ZOOLOGY OF NORTH AMERICA.

HAVING thus obtained a bird's-eye view of the physical features of North America, we will take a rapid survey of its zoölogy before we more minutely inspect the individuals of which it consists.

In a region of extent so vast as the continent of America, reaching from the Arctic Circle at one end far away towards the Antarctic Ocean at the other—with dense forests, under a tropical sun, in some parts ; open plains, lofty mountains, or a network of rivers and streams, vast lakes and marshes, in others—we shall find all varieties of form in the animal kingdom. This gives to its study an especial interest. While the larger number of its members are especially local, confined in narrow spaces between two streams, others range beyond 50° and 60° of latitude. The puma wanders across the plains of Patagonia, and ravages the flocks of the settlers on the western prairies of the United States. The reindeer feeds on the moss-covered moors of the Arctic islands, and is chased by the hunters far south among the defiles of the Rocky Mountains. Vast herds of bison darken the plains of New Mexico, and reach the upper waters of the Saskatchewan.

The same wild fowl which hatch their young among the ice-surrounded cliffs of Northern Greenland are found sporting in the lakes of Central America; while some of the smallest of the feathered tribes, the gem-like humming-birds, have been seen flitting through the damp mists of Tierra del Fuego, sipping the sweets of Alpine flowers high up amid lofty peaks of the Andes, and appearing on the hill-sides in sight of Lake Winnepeg, on the north of Rupert's Land.

However, as we proceed in our survey, we shall be able to note such, and many other interesting facts connected with the zoology of the districts we visit.

We shall find in the northern portion of the continent, extending nearly as far south as the sixtieth degree of latitude, and even beyond that parallel, several animals which are identical with those inhabiting the same latitudes in Europe and Asia. The Polar or white bear, the sovereign of the Arctic world, ranges entirely round the Circle; and makes his way across the icy seas over the rugged snow-clothed rocks, so that he belongs as much to Europe and Asia as to America. The cunning wolverene, the ermine, the pine marten, the Arctic fox and common weasel, also inhabit the same latitudes of the three continents. Among the herbivorous quadrupeds, there are several which have made their way across the frozen ocean. The American elk, though called the moose, is identical with the same animal found in Asia and Europe; so is the reindeer, known here as the cariboo. Both, indeed, are Arctic animals, though they migrate to southern latitudes when the severer cold and depth of snow prevents them from obtaining the moss and lichens on which they feed. The little Polar hare ranges round the Arctic Circle; but there is one animal, the musk-ox, which, being truly an Arctic quadruped, is unknown either in Asia or Europe, and therefore belongs exclusively to America.

Of the feathered tribes, the larger number of individuals, as might be supposed, are common to the northern portions of the three continents. Among these are the golden eagle, the white-headed or sea eagle, the osprey, the peregrine falcon, the gyrfalcon, the merlin goshawk, the common buzzard, rough-legged buzzard, hen-harrier, long-eared owl, short-eared owl, great snowy owl, and Tengmalm's owl. Nearly all the ducks and other swimming families, as might be expected, are also identical, as they can make their way with ease round the Circle, and find the same food and conditions of life. The waders, however, are generally distinct from those of Europe, as are the grouse inhabiting the same parallels of latitude. Only one or two have been found in Europe, as well as in America.

We must now take a glance at the animals which are distinctly American. In the first place, there are three bears —the savage grizzly of the Rocky Mountains; the cunning black bear; and the bear of the Barren Grounds. The beaver might take the first rank among American animals, for his sagacity, if not for his size. Then comes the Canada otter; the vison or minx; the clever little tree-loving raccoon; the American badger, differing from his European relative; and the pekan. There are several varieties of wolves, differing in size and somewhat in habits, but all equally voracious. There are several species of foxes, and no less than thirty of lem- mings, marmots, and squirrels, all of which are to be found within the more northern latitudes of the New World. There are three hares—known as the American, the prairie, and the little chief hares—which range over the northern continent.

Of the large animals we have the wapiti, a species of deer; two species of the black-tailed deer; a long-tailed deer; and the prong-horned antelope; also the wild goat; the big-horn sheep of the Rocky Mountains; and last, though not least, the American bison, familiarly known as the buffalo—the inhabitant of the wide-spreading plains and prairies extending from the Arctic Circle to Mexico.

Among the land birds, especially the birds of prey, there are several which are spread over the greater part of the northern continent, some indeed being found also in great numbers in South America. These are the turkey vulture, the black vulture, the little rusty-crowned falcon, the pigeon hawk, slate-coloured hawk, red-tailed buzzard, American horned owl, little American owl, and five other species of falcons. The perchers are less widely distributed.

There are, however, numerous families of insectivorous birds peculiar to America, which either permanently inhabit the more genial portions of the continent, or pay annual visits to those regions where the richest fruits abound and insect life prevails, affording them an abundant banquet. These migrating birds, as the winter draws on, take their departure southward to the warmer climate of Mexico, where they find abundance of food. As the summer returns, and the fruits of the orchard, the corn of the field, and wild berries ripen, and insects increase in numbers, vast flocks of warblers, woodpeckers, maize-birds, fly-catchers, thrushes, hang-nests, pigeons, blue-birds, and others return from their southern pilgrimage, to feed on the minute creatures which now people the plains, the hill-sides and forests, and on the abundant productions of the earth, enlivening the forests with their varied plumage, and delighting man by their melodious notes.

The number of gallinaceous birds is extremely limited. America can, however, boast of its native wild turkey—one of the most magnificent game-birds in existence. There is also the pinnated or Cupid Grouse. The Barren Grounds of Kentucky, and a few other districts, are inhabited by the ruffle grouse, which is also often called the pheasant. It ranges to a considerable distance northward, and Dr. Richardson found it even on the borders of the Polar regions. There is likewise a small-sized partridge, which is improperly called the quail.

With the exception of the golden plover, few of the wading birds resemble those of Europe. The snipe, the woodcock, the curlew, most of the sand-pipers, together with the coot and the water-hen, are distinct from those of Europe, and are not only peculiar to America, but few of them have been found to the south of the line. One of the most magnificent birds is the American flamingo, which is of a more beautiful and intense scarlet than that of Europe, and fully as tall; another bird, the wood-ibis, has the same form as the glossy ibis of southern Europe. In Carolina and Florida is found the magnificent scarlet ibis, but it seldom makes its way to the northern parts of the Union. There are several large and beautiful species of herons. Although most of the duck tribe range throughout the continent, there are some—such as the summer or tree duck of South Carolina—which range from the States to the warmer shores of the southern provinces, while the celebrated canvas-back duck, so highly prized at table, is found chiefly in the temperate parts of the continent. The rest of the duck tribe inhabit the northern regions, only quitting them for the United States during the severity of winter.

CHAPTER V.

THE MOOSE, OR ELK.

WE shall not introduce the animals we are about to inspect according to a systematic classification, but bring them forward as they appear to the eye of the traveller or sportsman, giving the largest and the most important the first place. Our object is rather to view the characteristic animals of each region we visit than to attempt a scientific examination of the whole animated kingdom of the world—a task which must be left to those who have far more time at their disposal than we possess.

We will begin, therefore, with the animals belonging to the ruminantia—the eighth in natural order; taking next the carnivora—the fifth; and the smaller rodentia—the sixth; while the birds and reptiles will follow in due course. Among these, however, we shall select only the most notable and curious; for although North America does not teem with animal life in the same degree as the southern half of the continent, were we to attempt to introduce all those existing in it we could give but a meagre account of each.

Without further preface, therefore, we will commence our survey with the elk.

The monarch of the American pine forests—the superb moose or elk—ranges from the mouth of the Mackenzie River to the shores of the Atlantic, at the eastern extremity of Nova Scotia, and passing the great lake region, is found even as far as the State of New York. Observe him as he stands with huge palmated horns ready for action, his vast nostrils snuffing up the scent coming from afar; his eyes dilated, and

THE MOOSE, OR ELK.

ears moving, watching for a foe; his bristly mane erect; his large body supported on his somewhat thick but agile limbs, standing fully six feet six inches in height at the shoulder, above which rise the head and antlers. The creature's muzzle is very broad, protruding, and covered with hair, except a small moist, naked spot in front of the nostrils. He has a short, thick neck, the hair thick and brittle. The throat is somewhat maned in both sexes. So large is the cavity of the nose, that a man may thrust his arm right into it. The intermaxillaries are very long, and the nasals short. He differs from the European elk only by having much darker hair,—the coat of the male, when in its prime, at the close of the summer, being completely black. Under the throat the males have a fleshy appendage termed the bell, from which grow long black hairs. The bristles on his thick muzzle are of a lighter colour than those of the coat, being somewhat of a reddish hue. The neck and shoulders are covered with very fine soft wool, curiously interwoven with the hair. Out of this the Indians manufacture soft, warm gloves. The moose hair is very brittle and inelastic. It is dyed by the Indians, and employed for ornamenting numerous articles of birch bark. The moose is of cautious and retiring habits, generally taking up his abode amid the mossy swamps found round the margins of the lakes, and which occupy the low ground in every direction. Here the cinnamon fern grows luxuriantly, while a few swamp maple saplings and mountain ash trees occur at intervals, and afford sufficient food to the moose.

It is to these regions the bull retires with his consort, and remains for weeks together, claiming to be the monarch of the swamp ; and should he hear the approach of a distant rival, he will crash with his antlers against the tree stems, making sudden mad rushes through the bushes, the sound of his blows reverberating to a distance. He has also a curious custom of tearing up the moss over a considerable area, exposing the black mud by pawing with the fore-feet. He continually visits these hills, and in consequence a strong musky effluvia arises from them. The Indian hunter, by examining them, can ascertain without fail when they were last visited by the animal. He utters loud sounds both by day and night, described by the Indians in their guttural voices as "quoth, quoth," but occasionally becoming sharper and more like a bellow when he hears a distant cow. The cow utters a prolonged and strangely wild call. This is imitated by the Indian hunter through a trumpet composed of rolled-up birch bark, when his dogs are in chase of the animal; and the bull being by this means attracted towards him, becomes more easily his victim. .

A WOUNDED ELK.

During the early part of the year, and the summer, the antlers are growing; but this process ceases early in September, when the moose has got rid of the last ragged strip of the deciduous skin against the young larch-trees and alder-bushes. He now stands ready to assert his claims against all rivals. At this season the bulls fight desperately; often the collision of the antlers of huge rivals, driven with mighty force by their immense and compact necks, is heard to a great distance, like the report of a gun on a still autumnal evening. They probably approach from different directions, regardless of the rugged ground, the rocks, and fallen trees in their course, bellowing loudly, and tearing up the ground with their horns. Now they catch sight of each other, and rush together like two gladiators. Now butting for some time till their antlers become interlocked, perhaps both fall struggling to the ground. Frequently portions of skeletons, the skulls united by firmly-locked antlers, have been found in some wilderness arena, where a deadly fight has occurred. A magnificent pair of horns thus interlocked is to be seen in the Museum of the Royal College of Surgeons. Terrible must have been the fate of the combatants, illustrating Byron's lines :—

> " Friends meet to part ;
> Love laughs at faith :
> True foes once met,
> Are joined till death."

Captain Hardy says he has twice heard the strange sound emitted by the moose, which, till he became acquainted with its origin, was almost appalling. It is a deep, hoarse, and prolonged bellow, more resembling a feline than a bovine roar. Sometimes the ear of the hunter is assailed by a tremendous clatter from some distant swamp or burned wood. It is the moose, defiantly sweeping the forest of pines right and left among the brittle branches of the ram pikes, as the scaled pines hardened by fire are locally termed. When, however, the moose wishes to beat a retreat in silence, his suspicions being aroused, he effects the process with marvellous stealth. Not a branch is heard to snap, and the horns are so carefully carried through the densest thickets, that a rabbit would

make as much noise when alarmed.   He will also, when hard
pressed, take the most desperate leaps to avoid his foes.

A DESPERATE LEAP.

Though he seldom or never attacks human beings ·when
unassailed, he will do so occasionally when badly wounded,

if nearly approached. An old Indian hunter had one day followed up a moose, and wounded the animal, when it turned on him. There being no tree near, he jammed himself for safety between two large granite boulders which were at hand. The aperture, however, did not extend far enough back to enable him to get altogether out of the reach of the infuriated bull, which set on him with its fore-feet, and pounded him so severely that several of his ribs were broken; indeed, for several years afterwards he was nearly bent double by the severe beating he had received.

In the summer, when the plague of flies commences, the moose takes to the water to avoid their bites. There are several species—one termed the moose-fly--which are equally annoying to the hunter. The animal strives to free himself from their irritation by running among bushes and brambles; and should he reach a lake, he will plunge into the water, allowing only his nostrils and mouth to remain above the surface. Sometimes, indeed, he will dive altogether, and is frequently known to hide himself from his pursuers by remaining for a long time below the water. He also feeds upon the tendrils and shoots of the yellow pond-lily, by reaching for them under water. An Indian, on one occasion, was following the track of a moose, when it led him to the edge of a little round pond in the woods, whence he could find no exit of the trail. After waiting for some time, he beheld the head of the animal rising above the surface in the very middle of the pond. While hastening for his gun, which he had left at a little distance, the moose made for the opposite shore, and emerging from the water, regained the shelter of the forest ere he could get round for a shot. The animals have been known also to visit the sea-shore, and one was seen swimming off to an island over a mile distant, which he reached in safety.

The moose feeds chiefly on the leaves of young shoots and bushes, or the smaller trees—the red and other maples, the white birch, the balsam, fur, poplar, and mountain ash ; and occasionally, as has been said, on the roots of the yellow pond-lily, with a bite now and then at a tussack of broad-leaved grass growing in the dried bogs. To get at the foliage beyond the reach of his muzzle, he frequently charges a young tree and rides it down, till he has brought the tempting leaves within his reach.

The horns of the animal begin to sprout in April, the old pair having fallen some time before. In the middle of this month the coat is shed, when the animal for some time afterwards presents a very rugged appearance. The cow towards the end of May produces one or two calves, generally near the margin of a lake, or in one of the densely-wooded islands, where they are secure from the attacks of the bull moose, who, cruel tyrant that he is, often destroys them. Rarely more than two are born at a time.

Besides its human foes, the moose is attacked occasionally by the bear. Captain Hardy describes coming upon the traces of a recent struggle between a young moose and one of these animals. "The bear had evidently stolen through the long grass upon the moose, and had taken him at a disadvantage in the treacherous bog. The grass was very much beaten down, and deep furrows in the soil below showed how energetically the unfortunate moose had striven to escape from his powerful assailant. There was a broad track plentifully strewed with moose hair, showing how the moose had struggled with the bear, to the wood, where, no doubt, the affair ended, and the bear dined."

what is called a yard; and in Canada, where its depth is very great, they have to remain in it during the whole winter, feeding round the area on the young wood of deciduous trees. In Nova Scotia, however, they migrate to other localities, when they have consumed the more tempting portions of food in the yard. In the morning and afternoon they are found feeding, or chewing the cud; but at noon, when they lie down, they are difficult to approach, as they are then on the alert, employing their wonderful faculties of scent and hearing to detect the faintest taint or sound in the air, which might indicate the approach of danger. The snapping of a little twig, the least collision of a rifle with a branch, or crunching of the snow under the mocassins, will suffice to arouse them. Curiously enough, however, they are not alarmed by any sound, even the loudest, to which they have been accustomed. The hunter has, therefore, to approach the yard with the greatest possible caution, in order to get a shot.

We will, however, start off on a moose hunt, in autumn, with a practical Indian hunter. The air of the autumnal night is frosty and bracing. The moose are moving rapidly from place to place. Night is drawing on. The last fluttering of the aspens dying away, leaves that perfect repose in the air which is so necessary to the sport. The moon rises, shedding a broad and silvery light through the forest. Mysterious sounds greet our ears. The Indian hunter is provided with his trumpet of birch bark, in the form of a cone, about two feet in length. He shelters himself behind the edge of the banks, a clump of bushes, or rocks; and now he emits the cry of the cow moose, so exactly, that the male animal is easily deceived by it. He waits: there is no response. An interval of fifteen minutes elapses; still no reply is heard. Again the Indian sends his wild cry pealing through the wood. Presently a low grunt, quickly repeated, comes from some distant hill; and the snapping of branches and falling trees attests the approach of the bull. The hunter is now doubly careful; kneeling down, and thrusting the mouth of his call into some bushes close by, he utters a lower and more plaintive sound. At length an answer reaches his ears. The snapping of the branches is resumed; and presently the moose is seen stalking into the middle of the moonlit "barren." Our weapons are ready; and as the magnificent animal stands looking eagerly around in the woodland amphitheatre, a rifle ball, laden with death, brings him to the ground.

In some districts the Indians employ another method of calling. They conceal themselves in a swamp, in the midst of some damp mossy valley, during a dark night. One holds a torch of birch bark with a match ready for lighting. The hunter calls, and the moose approaches more readily than towards the open "barren." When the creature is within distance of the deadly rifle, the match is applied to the torch, which, flaring up, illuminates the swamp, and discovers the startled moose standing amidst the trees, and incapable apparently of flight. The Indians declare that he is fascinated by the light; and though he may walk round and round it, he will not leave the spot, and thus presents an easy mark to the hunter's rifle.

Let us set forth on an expedition to "creep" moose, which

CAMP OF MOOSE-HUNTERS

may be described as a similar mode of hunting to stalking. The ground we select is among the " barrens " before described. It is strewed with dead trees in all directions, amid which briars and bushes have grown up, and conceal their sharp, broken limbs, and the rough granite rocks scattered in all directions. Here, collecting wood for burning, we form our camp, and sit round the blazing fire, on which a well-filled frying-pan is hissing, while we are covered by our blankets to protect ourselves from the pattering rain-drops. Our suppers over, we stretch ourselves for repose, and gradually fall asleep, as the snapping of the logs on the fire, the pattering of the rain, and the hootings of the owls in the distant forest become less and less distinct. Our Indian brings us notice in the morning that two moose have passed close to the camp during the night. However, in spite of the plaintive call from the treacherous bark trumpet, they will not approach, having been forewarned of danger by the smell of our camp-fire. We make our way amid the bushes, already leafless, except that here and there are seen bunches of dwarf maples with a few scarlet leaves of autumn still clinging to them. Presently our companion whispers, "Down—sink down! slow—like me!" A magnificent bull appears about five hundred yards off. The wind is blowing from him to us. The Indian utters the usual call ; but the moose does not answer, having already a companion close at hand. Presently he lies down in the bushes, and we worm ourselves slowly and laboriously towards the edge of the alder swamp. Gently lowering ourselves into the swamp, we creep noiselessly through the dense bushes, their thick foliage closing over our heads. It is an anxious moment!—the slightest snapping of a bough, the knocking of a gun-barrel against a stem, and the game is off.   "We must

bushes. At length, as we reach the edge of the swamp, the

" CREEPING " THE MOOSE.

great animal rises directly facing us, gazing steadily towards us. We fire. A headlong stagger follows the report; and the creature, turning round, is hidden from sight behind a clump of bushes. The Indian at the same time fires at a large cow moose who has, unknown to us, been lying close to the bull. We dash forward a few paces. On the other side the great bull suddenly rises in front of us and strides on into thicker covert. Another shot, and he sinks lifeless at our feet.

THE CARIBOO, OR REINDEER.

We have before mentioned the extensive tracts existing in North America, which, from their desolate appearance, are appropriately called " Barrens." Far as the eye can reach the whole ground is seen strewn with boulders of rock and fallen trees, scattered round in the wildest confusion. Here and there charred stumps rise from the green-sward; in some spots clumps of spruce are seen, against which the white stems of the graceful birch stand out in bold relief; while the bank of some stream, or the margin of a lake, is marked by fringing thickets of alder. In many parts are moist, swampy bogs, into which the sportsman sinks ankle-deep at every step. The ground, however, is everywhere thickly carpeted by a luxuriant growth of a species of lichen. It possesses wonderfully nutritive qualities; so much so, that large quantities of alcohol have lately been extracted from it, as well as from other lichens growing in sub-arctic regions. It is the chief food of the cariboo, which animal frequents these desolate-looking "barrens."

Visiting one of these "barrens," we may perchance fall in with several of the noble-looking animals known in Europe and Asia as the reindeer, though we must look sharp to recognize them; for so similar are they in colour to the rocks and general features of the ground, that only the keen eye of the Indian can easily detect them, especially when they are lying down. Should we approach them on the weather-side, or should the slightest noise be made, they will quickly detect us. Up they spring, and after a brief stare, make off in graceful bounds at a rapid rate. Now, having got beyond danger, they drop into a long swinging trot, and proceed in single file across the "barren," till they enter the line of forest in the far distance.

The cariboo of North America is a strongly-built, thick-set animal, compared to the more graceful of his relatives. He carries on his head a pair of magnificent antlers, varying greatly in different specimens—some palmating towards the upper ends, others with branches springing from the palmated portions. In most instances there is but one developed brow antler, the other being a solitary curved prong. The back of the cariboo is covered with brownish hair, the tips of which are of a rich dun gray, whiter on the neck than elsewhere. The nose, ears, and outer surface of the legs and shoulders are of a brown hue. The neck and throat are covered with long, dullish white hair, and there is a faint whitish patch on the side of the shoulders. The rump and tail are snowy white, while a band of white runs round all the legs, joining the hoofs.

As winter approaches, the hair grows long, and lightens considerably in hue. Frequently, indeed, individuals may be seen in a herd with coats of the palest fawn colour—almost white. The muzzle is entirely covered with hair. The fur is brittle, and though in summer it is short, in winter it is longer and whiter, especially about the throat. The hoofs are broad, depressed, and bent in at the tip. The full-grown bucks shed their horns, and it is seldom that they are seen in a herd after Christmas. The female reindeer, however, retains hers during winter. Several theories have been advanced to account for this. There seems no doubt, however, that the object is to enable the female to protect her fawns from the males, who are apt to attack the young and destroy them.

The cariboo is gregarious, and males, females, and young herd together at all seasons; and by this provision of Nature the females are able to defend the young, who would otherwise be subjected to injury. In another respect these animals are wonderfully provided for the mode of existence they are compelled to pursue. Not only have they to cross wide snow-covered districts, but frequently to pass across frozen

expanses of water.   To enable them to do this in the winter,
the frog of the foot is almost entirely absorbed, and the edges
of the hoof, now quite concave, grow out in sharp ridges,
each division on the under surface presenting the appearance
of a huge mussel-shell, and serving the office of natural
skates.   So rapidly does the shell increase, that the frog does
not fill up again till spring, when the antlers bud out.   With
this singular conformation of the foot, it has a lateral spread;
and an additional assistance for maintaining a foothold on
slippery surfaces is given by numerous long, stiff bristles
which grow downward at the fetlock, curving over entirely
between the divisions.   The cariboo is thus enabled to proceed
over the snow, to cross frozen lakes, or ascend icy precipices,
with an ease which places him, when in flight, beyond the
reach of all enemies, except perhaps the nimble and untiring
wolf.

The cariboo is essentially a migratory animal.   There are
two well-defined periods of migration, in the spring and
autumn.   Throughout the winter it appears also seized with
an unconquerable desire to change its residence.   One day it
may be found feeding quietly through the forests in little
bands, and the next, perhaps, all tracks show a general move
in a certain direction.   The animals join the main herd after
a while, and entirely leaving the district, travel toward new
feeding-grounds.   Though often found in the same woodlands
as the moose, they do not enjoy each other's company.   In
severe winters the cariboos travel to the southernmost limits
of their haunts, and even sometimes enter the settlements.
Not being aquatic, like the moose, to avoid the flies in
summer they ascend the mountain ranges, where they can
be free from their attacks.   The hunter, however, follows

them, and their speed being of no avail among the precipices,

many are shot. During most of the year the flesh of the animal is dry and tasteless ; but it possesses a layer of fat, two or more inches thick, which is greatly esteemed. This, with the marrow, is pounded together with the dried flesh, and makes the best kind of pemmican—a food of the greatest value to the hunter. The cariboo lives in herds, sometimes only of ten or twenty, but at others consisting of thirty or more individuals. They range across the whole width of the continent, being found in great numbers to the west of the Rocky Mountains, especially at the northern end of British Columbia. Although specifically identical with the reindeer of Europe, it has never yet been trained by Indians or Esquimaux to carry their goods or draw their sleighs, as in Lapland and along the Arctic shores of Asia.

SHOOTING THE CARIBOO.

THE WAPITI, OR CANADIAN STAG.

In the wilder parts of the Southern States of the Union, herds of the magnificent Canadian stag or wapiti—popularly called the elk—range amid the woods and over the prairies. Sometimes three or four hundred are found in one herd, always

led by an old buck, who exacts from them the strictest obedi-
ence—compelling them to halt or move onward as he judges
necessary.   Now the superb herd of long-horned creatures are
seen to wheel to the right or left, now to advance or retreat
at the signal he issues.

The wapiti is indeed a grand animal, growing to the height
of the tallest ox, and endowed with wonderful activity, as
well as power.
See him as he
dashes through
the forest, his
branched horns
separating in ser-
pentine curves, six
feet from tip to
tip, laid close over
his back as he
makes his way
amid the trees.
His head is of a
lively, yellowish-
brown hue, the
neck covered with

THE WAPITI.

reddish and black hairs, the latter of considerable length,
descending in a thick bunch below it.   They are among the
fiercest of the deer tribe.   The bucks often enter into desperate contests with each other, battling with their huge
horns—the fight frequently ending only with the death of the
weaker rival.   Sometimes their horns have become so inex-
tricably interlocked, that both have fallen to the ground, and,
unable to rise, have perished miserably.   They will frequently,

SHOOTING THE CARIBOO.

when wounded, attack their human assailants; and the bold hunter, if thus exposed with rifle unloaded to their fierce assaults, will rue the day his weapon failed to kill the enraged quarry at the first shot.

The wapiti, when pursued, will boldly plunge into the

LAKE-HUNTING.

lake or broad river, and breast the rapid current to avoid his foes; or will occasionally, if hard pressed, attack the bold hunter who ventures to follow in his light canoe.

His cry is a sharp whistling sound, which rings through the air far and wide on a calm day.   He feeds on the branches of the trees and grass, and in winter scrapes, with his powerful fore-feet, deep into the snow, to obtain the lichens and dry herbage which grow beneath.   His flesh for several months in the year is dry and coarse, but his hide is much prized by the Indians, who manufacture from it a leather of a peculiarly soft character, which retains that quality after being wet,—instead of turning hard, as is the case with that manufactured from other deerskins.   A remarkable feature of the wapiti is that the horns differ in form almost as greatly as do those of the branches of trees, no two specimens being found with them exactly alike.

THE KARJACOU, OR VIRGINIAN DEER.

The most graceful of the deer tribe, the karjacou, scours in large herds across the prairies, frequently entering the haunts of man.   Yet so easily is it scared that it takes to flight at the very appearance of a human being.   Curiously enough, however, it will again return to its favourite feeding-grounds, even though the hunter's rifle may lay low many of the herd. It is about the size of the fallow-deer, and of a light brown hue.   Its horns are slender, and have numerous branches on the interior sides, but are destitute of brow antlers.

Let us watch a herd startled by our approach.   Away they spring, leaping into the air, turning their heads in every direction to ascertain the cause of their alarm, and then rush off at full speed ; but in a short time, if they are not followed, we may see them return, especially as night draws on, and crouch down in their accustomed sleeping-places.   Should a salt lake be near, they will come in vast numbers to lick up with their tongues the saline particles adhering to the surrounding stones, where the salt has crystallized from the evaporation of the water.

They are at all times thirsty, and they require constant draughts of pure water, to obtain which they are sure to visit the nearest stream or spring as night is about to close over the scene. Wherever the tenderest herbage grows upon the plain, there the karjacou comes to crop it during summer. In winter he finds an abundant supply of food from the buds and berries, or fallen fruits; or, when snow is on the ground, he eats the string moss hanging in masses from the trees. He willingly takes to the water, and will cross a lake or broad river, swimming at a rapid rate with his whole body submerged, his head alone appearing above the surface; thus he will often baffle his pursuers, even though they may follow him with a boat. He has been known, indeed, when hard pressed near the sea-coast, to plunge into the ocean, and buffeting the waves, to make his way far from the land, rather than be captured.

His flesh affords the Indian a large portion of his winter supply of food, while his skin is manufactured into clothing, the leather from it being especially soft and pliable. From the settlers in the western provinces he receives little mercy, as, without hesitation, he leaps their fences, banqueting on their growing corn or vegetables; and, after doing all the mischief in his power, by his activity generally again makes his escape. No animal surpasses in beauty the young fawn, the fur of which is of a ruddy brown tint, ornamented with white spots arranged in irregular lines, merging occasionally into wide stripes.

Like others of his tribe, the male is excessively combative when meeting others of his own species; and a story is told of three animals thus encountering each other in a desert, when all their horns becoming entangled, they remained fixed, unable to separate, till they sank together on the ground, their skulls and skeletons afterwards being discovered, thus giving evidence of the combat and its fatal result.

THE ANTELOPE.

No animal of the American wilds surpasses the antelope in beauty. The little creatures congregate in herds of many

THE ANTELOPE.

thousands, though, from the exterminating war waged against them by the Indians, they have greatly decreased in numbers.

The size of the antelope is about that of the common red-deer doe ; the colour somewhat between buff and fawn, shaded here and there into reddish-brown, and a patch of pure white on the hind-quarters. This gives rise to the expression of the hunter, when he sees it flying before him, that the creature is "showing its clean linen." The ears are placed far back on the head, are very long, and curved so much that at a distance they appear like horns, while the horns themselves appear as if coming out of the animal's eyes ; they are long and slender, curving slightly backwards, and have no branches, except a little bud, which is developed when the creature is about two years old. The chief peculiarity of the animal is its lack of a dewlap.

The feet have no rudimentary hoofs like the deer, yet this want in no way interferes with its speed. Often the creature may be seen for a moment browsing not fifty yards off, the next it has dwindled to a mere speck, and is in another lost to sight. They do not leap like deer, but run with level backs, as sheep do, their legs glancing faster than sight can follow. In vain the hunter attempts to follow the rapid movements of the creatures on horseback. Perhaps they will let him approach to within a short distance, and then away they float on a line at right-angles to their former retreat. To come up with them, indeed, as an American writer observes, is as hopeful an undertaking as trying to run down a telegraphic message. The only way to get near them is by a stratagem. They are not afraid of horses, and the hunter, by walking behind his horse, may often approach a herd without being discovered, provided the wind blows from them. He then pickets his horse with a sharp stake, and sinking down in the grass he ties a bright-coloured handkerchief to the end of his ramrod ; he then crawls forward on hands and knees, dragging his rifle, till he approaches still nearer, when he remains concealed, and lifts his flag in the air. The antelopes, on catching sight of it, stop browsing, and raising their heads, peer towards it, exhibiting no signs of fear. For a moment he drops his flag ; the beautiful creatures then resume their repast, but their curiosity gets the better of their prudence.

Again they look forward, when the flag is once more raised and waved slowly backward and forward. The antelopes have now their curiosity excited to the utmost ; for a moment they stop irresolute, then advance a few steps snuffing the air. Once more the flag sinks out of sight ; they seem to be asking each other what is the cause of the strange sight they have seen. Again it is raised ; they draw nearer and nearer, till they are within range of the hunter's deadly rifle ; he fires, and almost to a certainty one of the beautiful animals springs into the air and tumbles head-foremost on the ground. For a moment the survivors run off from their fallen friend, but seldom go far. Once more they return within easy rifle-shot of the hunter. Unless, however, he requires the meat, he must be greatly lacking in right feeling if he slaughters use-lessly so beautiful an animal. The antelope becomes so easily confused, that when met on the prairies it frequently runs headlong into the midst of the travellers. The creatures are often killed by being surrounded, when the whole herd are driven into an enclosed spot and become the easy prey of the hungry hunters.

THE BIGHORN, OR MOUNTAIN SHEEP.

Amid the almost inaccessible peaks of the Rocky Mountains, herds of animals with enormous horns may be seen leaping from rock to rock, sometimes descending at one spring from a height of twenty or thirty feet—when, the Indians assert, they invariably alight on their horns, and by this means save their bones from certain dislocation. They are bighorns, or moun-tain sheep, and are considered the chief game of these regions. The animals appear to partake both of the nature of the deer and of the goat. They resemble the latter more especially in their habits, and in frequenting the most lofty and inaccessible regions, whence, except in the severest weather, they seldom descend to the upland valleys.   In size the bighorn is between the domestic sheep and the common red-deer of America, but is more strongly built than the latter.   It is of a brownish-dun colour, with a somewhat white streak on the hind-quarters. The tail is shorter than that of the deer, and tipped with black.   As the age of the animal increases, the coat becomes of a darker tinge.   The horns, of the male especially, are of great size, curving backwards about three feet in length, and twenty inches in circumference at the roots.

Frequently on the highest spot one of the band is stationed as a sentinel, and whilst the others are feeding he looks out for the approach of danger.   They have even more acute sight and smell than the deer.   On an alarm being given the whole herd scampers up the mountain, higher and higher, every now and then halting on some overhanging crag and looking down on the object which may have caused them alarm ; then once more they pursue their ascent, and as they bound up the steep sides of the mountains throw down an avalanche of rocks and stones.

Occasionally the young lambs are caught and domesticated by the hunters in their mountain homes, when they become greatly attached to their masters, amusing them by their merry gambols and playful tricks.   Attempts have been made to transport them to the States; but although milch-goats have been brought to feed the lambs, they have suffered by the change from the pure air of the mountains to the plains, or they have not taken kindly to their foster-mothers, and have invariably perished on the journey.

The creatures reach a height of three feet six inches at the shoulders, while the horns are of about the same length. In colour they vary greatly, changing according to the season of the year.

THE BISON, COMMONLY CALLED THE BUFFALO IN AMERICA.

Throughout the wide-extending prairies of North America, from north to south to the east of the Rocky Mountains, vast herds of huge animals—with shaggy coats and manes which hang down over the head and shoulders reaching to the ground, and short curling horns, giving their countenances a ferocious aspect—range up and down, sometimes amounting to ten thousand head in one herd. They commonly go by the name

THE BISON.

of buffaloes, but are properly called bisons. Clothed in a dense coat of long woolly hair, the buffalo is well constituted to stand the heats of summer as well as the cold of the snowy plains in the northern regions to which he extends his wanderings.

Let us look at him as he stands facing us on his native

plains, his red eyes glowing like coals of fire from amid the mass of dark brown or black hair which hangs over his head and neck and the whole fore part of his body. A beard descends from the lower jaw to the knee; another huge bunch of matted hair rises from the top of his head, almost concealing his thick, short, pointed horns standing wide apart from each other. As he turns round we shall see that a large oblong hump rises on his back, diminishing in height towards the tail : that member is short, with a tuft of hair at the tip. The hinder part of the body is clothed with hair of more moderate length, especially in summer, when it becomes fine and smooth, and soft as velvet. From his awkward, heavy appearance, when seen at a distance, it would not be supposed that he is extremely active, capable of moving at a rapid rate, and of continuing his headlong career for an immense distance. So sure of foot is he, also, that he will pass over ground where no horse could follow, his limbs being in reality slender, and his body far more finely proportioned than would be supposed till it is seen stripped of its thick coating of hair. While his thick coat protects him from the cold, he is also provided with a broad, strong, and tough nose, with which he can shovel away the snow and lay bare the grass on which he feeds. Sometimes, however, when a slight thaw has occurred, and a thin cake of ice has been formed over the snow, his nose gets sadly cut, and is often seen bleeding from the effects of his labours. It is said that when a herd comes near the settlements, the domesticated calves, and even the horses, will follow the buffalo tracks, and graze on the herbage which they have disclosed and left unconsumed.

The flesh of the buffalo, especially that of the cow, is juicy, and tender in the extreme. The most esteemed portion is

HERD OF BUFFALOES CROSSING A STREAM

that composing the hump on its back, which gives it so strange an aspect. It is indeed frequently killed merely for the sake of this hump, and the tongue and marrow-bones. Sometimes, also, when parched with thirst, the hunter kills a buffalo to obtain the water contained within certain honey-combed cells in its stomach. The buffalo is provided with this reservoir, in which a large quantity of pure water can be stored, that it may traverse, without the necessity of drinking, the wide barren plains where none can be obtained. Vast numbers, without even these objects in view, are wantonly slaughtered, and the chief part of the flesh utterly wasted, by the thoughtless Indians of the plain, who have thereby deprived themselves of their future support. Many tribes depend almost entirely for their subsistence on the buffalo, of which the flesh is prepared in several ways. When cut up into long strips, and dried in the sun till it becomes black and hard, it will keep for a long time. It is also pounded with the fat of the animal, and converted into *pemmican*—an especially nutritious food, which, if kept dry, will continue in good order for several years.

The prairie Indians make use of the hide for many purposes. They scrape off the hair and tan it, when it serves them for coverings for their tents. It is also carefully dressed, when it becomes soft and impervious to water. It is then used for clothing. Some of the tribes also form their shields from it. The hide is pegged down on the ground, when it is covered with a kind of glue. In this state it greatly shrinks and thickens, and becomes sufficiently hard to resist an arrow, and even to turn aside an ordinary bullet which does not strike directly.

The buffalo is especially a gregarious animal, and is found

in herds of immense size, many thousands in number. Their dark forms may often be seen extending over the prairie as far as the eye can reach, a mighty moving mass of life. Onward they rush, moved by some sudden impulse, making the ground tremble under their feet, while their course may be traced by the vast cloud of dust which floats over them as they sweep across the plain. They are invariably followed by flocks of wolves, who pounce on any young or sick members of the herd which may be left behind. They range throughout the whole prairie country, from the " Fertile Belt," which extends from the Red River settlement to the Rocky Mountains in British Central America, to Mexico in the south. The bulls are at times excessively savage. They often quarrel among themselves, and then, falling out of the herd, they engage in furious combats, greatly to the advantage of the pursuing wolves. In the summer, the buffalo delights in wallowing in mud. Reaching some marshy spot, he throws himself down, and works away till he excavates a mud-hole in the soil. The water from the surrounding ground rapidly drains into this, and covers him up, thus freeing him from the stings of the gnats and flies which swarm in that season.

The buffalo is hunted on horseback both by whites and by Indians, though the sport is one in which a considerable amount of danger must be braved. Let us set off from a farm in the Western States, on the border of the prairie. We have one or two nights to camp out before we reach the buffalo grounds. Mounting our horses by break of day, after an early breakfast, we ride on with the wind in our faces, and at length discover across the plain a number of dark objects moving slowly. They are buffaloes, feeding as they go. We see through our field-glasses that there are calves among them.

It is proposed that some of our party should ride round, so as to stampede the herd back towards us, and thus, by dividing them, enable us to get in the centre. We wait for some time, when we see a vast mass of hairy monsters come tearing over a hill towards us. We have shot several of the bulls, but our object is to secure their calves and cows. As the herd approaches us, it swings round its front at right-angles, and makes off westward. We dash forward, and divide it into two parties. We also separate, some of our hunters following one part of the herd, the others the remainder. The enthusiasm of our horses equals our own. Away we go ; nothing stops us. Now we plunge with headlong bounds down bluffs of caving sands fifty feet high,—while the buffaloes, crazy with terror, are scrambling half-way up the opposite side. Now we are on the very haunches of our game ; now before us appears a slippery buffalo wallow. We see it just in time to leap clear, but the next instant we are in the middle of one. Our horses, with frantic plunges, scramble out ; and on we go. We get closer and closer to the buffaloes, when a loud thundering of trampling hoofs sounds behind us. Looking over our shoulders, there, in plain sight, appears another herd, tearing down on our rear. For nearly a mile in width stretches a line of angry faces, a rolling surf of wind-blown hair, a row of quivering lights burning with a reddish-brown hue—the eyes of the infuriated animals. Should our horses stumble, our fate will be sealed. It is certain death to be involved in the herd. So is it to turn back. In an instant we should be trampled and gored to death. Our only hope is to ride steadily in the line of the stampede, till we can insinuate ourselves laterally, and break out through the side of the herd. Yet the hope of doing so is but small.

INDIANS HUNTING BUFFALO

On we rush rapidly as before, when suddenly, to our great satisfaction, the herd before us divides into two columns, to pass round a low hill in front. Still on we go, pushing our horses up the height. We reach the summit, the horses panting fearfully, and the moisture trickling in streams from their sides. But now the rear column comes on. They see us, not fifty rods off, but happily pay no attention to us. We dismount, facing the furious creatures. Should they not divide, but come over the hill, in a few moments we must be trampled to death. The herd approaches to within a hundred yards of the hill. We lift our rifles and deliver a couple of steadily aimed bullets at the fore-shoulders of the nearest bulls. One gives a wild jump, and limps on with three legs ; the other seems at first unhurt ; but just as they reach the foot of the mound, they both fall down. The whole host are rushing over them. We rapidly reload. The fate of their comrades, however, sends a panic into the hearts of the herd. Another falls just when they are so close that we could have sprung on their backs. At that moment they divide, and the next we are standing on a desert island, a sea of billowing backs flowing round on either side in a half-mile current of crazy buffaloes. The herd is fully five minutes in passing us. We watch them as they come, and as the last laggers pant by the mound we look westward and see the stampeders halting. We soon understand the cause. They have come up with the main herd. Yes, there, in full sight of us, is the buffalo army, extending its deep line as far as the western horizon. The whole earth is black with them. From a point a mile in front of us, their rear line extends on the north to the bluffs bounding the banks of the river on which we had camped. On the south it reaches the summits of some distant heights fully six miles away. When it is known that with our field-glasses we can recognize an object the size of a buffalo ten miles distant, and that the mass extends even beyond the horizon, some idea may be formed of the immense number of animals congregated in the herd. To say that there are ten thousand, would be to give a very low estimate of their numbers.

The same writer from whose work the above is taken, describes an extraordinary instance of friendship exhibited by a buffalo bull for one of his comrades. (Generally speaking, the buffalo, even in the pairing season, will forsake the wounded cow, and the cow will not stay one moment to protect her hurt calf.) He was out hunting on one occasion, when, having been for some time unsuccessful, and being anxious to retrieve his character by bringing home some meat to camp, he caught sight of two fine buffalo bulls on a broad meadow on the opposite side of a stream. Dismounting from his horse, he took steady aim at the nearest buffalo, which was grazing with its haunches towards him. The ball broke the animal's right hip, and he plunged away on three legs, the other hanging useless. He leaped on his horse, put spurs to its flanks, and in three minutes was close on the bull's rear. To his astonishment, and the still greater surprise of the two old hunters who came after him, the unhurt bull stuck to his comrade's side without flinching. He fired another shot, which took effect in the lungs of the first buffalo. The second moved off for a moment, but instantly returned to his friend. The wounded buffalo became distressed, and slackened his pace. The unwounded one not only retarded his, but coming to the rear of his friend, stood, with his head down, offering battle. "Here indeed was devotion which had no instinct to inspire it. The sight was sublime! The hunters could no more have accepted the challenge of the brave creature, than they could have smitten Damon at the side of Pythias. The wounded buffalo ran on to the border of the next marsh, and, in attempting to cross, fell headlong down the steep bank, and never rose again. Not till that moment, when courage was useless, did the faithful creature consider his own safety in flight. The hunters took off their hats as he walked away, and gave three parting cheers as the gallant buffalo vanished beyond the fringing timber."

The half-breed hunters of Rupert's Land make two expeditions in the year in search of buffaloes—one in the middle of June, and the other in October. They divide into three bands, each taking a separate route, for the purpose of falling in with the herds of buffaloes. These bands are each accompanied by about five hundred carts, drawn by either an ox or a horse. They are curious vehicles, roughly formed with their own axes, and fastened together with wooden pins and leather thongs, not a nail being used. The tires of the wheels are made of buffalo hide, and put on wet. When they become dry, they shrink, and are so tight that they never fall off, and last as long as the cart holds together. The carts contain the women and children, and provisions, and are intended to bring back the spoils of the chase. Each is decorated with some flag, so that the hunters may recognize their own from a distance. They may be seen winding off in one wide line extending for miles, and accompanied by the hunters on horseback. These expeditions run the danger of being attacked by the Sioux Indians, who inhabit the prairies to the south. The camps are therefore well surrounded by scouts, for the purpose of reconnoitring either for enemies or buffaloes. If they see the latter, they make a signal by throwing up handfuls of dust; if the former, by running their horses to and fro.

Mr. Paul Kane, the Canadian artist, describes one of these expeditions which he joined. On their way they were visited by twelve Sioux chiefs, who came for the purpose of negotiating a permanent peace; but whilst smoking the pipe of peace in the council lodge, the dead·body of a half-breed, who had gone to a distance from the camp, was brought in newly scalped, and his death was at once attributed to the Sioux. Had not the older and more temperate half-breeds interfered, the young men would have destroyed the twelve chiefs on the spot: as it was, they were allowed to depart unharmed. Three days afterwards, however, the scouts were observed making the signal of enemies being in sight. Immediately a hundred of the best-mounted hastened to the spot, and concealing themselves behind the shelter of the bank of a stream, sent out two of their number as decoys, to expose themselves to the view of the Sioux. The latter, supposing them to be alone, rushed upon them; whereupon the concealed half-breeds sprang up and poured in a volley which brought down eight. The others escaped, though several must have been wounded.

Two small herds having been met with, of which several animals were killed, the scouts one morning brought in word that an immense herd of bulls was in advance about two miles off. They are known in the distance from the cows by their feeding singly, and being scattered over the plain,—whereas the cows keep together, for the purpose of protecting the calves, which are always kept in the centre of the herd.

We will start at daybreak with our friend, and a half-breed

as a guide. Six hours' hard riding brings us to within a quarter of a mile of the nearest herd. The main body stretches over the plains as far as the eye can reach, the wind blowing in our faces. We should have liked to have attacked them at once, but the guide will not hear of it, as it is contrary to the law of his tribe. We therefore shelter ourselves behind a mound, relieving our horses of their saddles to cool them. In about an hour one hundred and thirty hunters come up, every man loading his gun, looking to the priming, and examining the efficiency of his saddle-girths. The elder caution the less experienced not to shoot each other,—such accidents sometimes occurring. Each hunter then fills his mouth with bullets, which he drops into the gun without wadding; by this means loading more quickly, and being able to do so whilst his horse is at full speed. We slowly walk our horses towards the herd. Advancing about two hundred yards, the animals perceive us, and start off in the opposite direction, at the top of their speed. We now urge our horses to full gallop, and in twenty minutes are in the midst of the stamping long-haired herd. There cannot be less than four or five thousand in our immediate vicinity,—all bulls ; not a single cow amongst them. The scene now becomes one of intense excitement,—the huge bulls thundering over the plain in headlong confusion, while the fearless hunters ride recklessly in their midst, keeping up an incessant fire but a few yards from their victims. Upon the fall of each buffalo the hunter merely throws, close to it, some article of his apparel to denote his own prey, and then rushes on to another. The chase continues for about one hour, extending over an area of about six square miles, where may be seen the dead and dying buffaloes to the number of five hundred. In spite of his horsemanship, more than one hunter has been thrown from his steed, in consequence of the innumerable badger-holes in which the plains abound. Two others are carried back to camp insensible. We have just put a bullet through an enormous bull. He does not fall, but stops, facing us, pawing the earth, bellowing, and glaring savagely. The blood is streaming from his mouth, and it seems as if he must speedily drop. We watch him, admiring his ferocious aspect, combating with death. Suddenly he makes a dash towards us, and we have barely time to escape the charge ; when, reloading, we again fire, and he sinks to the ground.

The carts bring in the slaughtered animals to the camp, when the squaws set to work, aided by the men, to cut them up, and prepare them for drying and for making pemmican. The women are soon busily employed in cutting the flesh into slices, and in hanging them in the sun on poles. The dried meat is then pounded between two stones till the fibres separate. About fifty pounds of it is put into a bag of buffalo skin, with about forty pounds of melted fat, which, being mixed while hot, forms a hard and compact mass. Hence its name, in the Cree language, of pemmikon—*pemmi* signifying meat, and *kon* fat—usually, however, spelled pemmican. One pound of pemmican is considered equal to four pounds of ordinary meat,—and it keeps for years, perfectly good, exposed to any weather.

The prairie Indians obtain buffaloes by driving them into huge pounds, where they are slaughtered. The pounds, however, can only be made in the neighbourhood of forests, from whence the logs for their formation can be obtained. The pound consists of a circular fence about 130 feet broad. It is constructed of the trunks of trees laced together with withes. with outside supports about 5 feet high. At one side an entrance is left about 10 feet wide, with a deep trench across it, on the outside of which there is a strong trunk of a tree placed, about a foot from the ground. The animals, on being driven in, leap over this, clearing the trench, which of course prevents them from returning. From the entrance two rows of bushes or posts, which are called " dead men," diverge towards the direction from which the buffaloes are likely to come. They are placed from 20 feet to 50 feet apart, and the distance between the extremities of the two rows at their outer termination is nearly two miles. Behind each of these " dead men " an Indian is stationed, to prevent the buffaloes when passing up the avenue from breaking out. Meantime, the hunters, mounted on fleet horses, range the country to a distance of eighteen or twenty miles in search of a herd. The buffalo has an unaccountable propensity which makes him endeavour to cross in front of the hunter's horse. They will frequently, indeed, follow a horseman for miles in order to do so. He thus possesses an unfailing means, by a dexterous management of his horse, of conducting the animals into the trap prepared for them. The men also conceal themselves in hollows and depressions in the ground, so as to assist in turning the herd, should they attempt to escape in that direction. And now some three or four hundred head of shaggy monsters are driven to the expanded mouth of the avenue. The horsemen follow in their rear, and prevent them turning back. Meantime the Indians stationed behind the " dead men" rise, shaking their bows, yelling, and urging them on. Thus they proceed, madly rushing on, the passage growing narrower and narrower, while they, pressed together, are unable to see the danger ahead. The foremost at length reach the fatal ditch, and leaping over, enter the pound, the hold their robes before every orifice, till the whole herd is brought in. They then climb to the top of the fence, and the hunters, who have followed closely in the rear of the buffaloes, spear and shoot with bows and arrows or firearms at the bewildered animals, rapidly becoming frantic with fear and terror in the narrow limits of the pound. A dreadful scene of confusion and slaughter then ensues. The older animals toss the younger. The shouts and screams of the Indians rise above the roar of the bulls, the bellowing of the cows, and the moaning of the calves. The dying struggles of so many powerful animals crowded together, create a revolting scene, dreadful for its excess of cruelty and waste of life." [*]

In consequence of this wholesale and wanton destruction, the buffalo has greatly diminished; and the Indians agree in the belief that their people, in like manner, will decrease till none are left. It is computed that for many years past no less than 145,000 buffaloes have annually been killed in British territory; while on the great prairies claimed by the United States a still greater number have been slaughtered. In one year—1855—on the British side of the boundary, there were 20,000 robes of skins received at York Factory alone; and probably not fewer than 230,000 head of buffalo were slaughtered in the previous year. This number would have been sufficient to sustain a population of a quarter of a million. Yet so vast a number of the animals are left to rot on the ground, that in all probability not more than

[*] Blind.

30,000 Indians. fed on the flesh of the slaughtered buffaloes.

The civilized fur-traders, however, with greater forethought, take means to preserve the flesh of the animals they kill in the neighbourhood of the forts, so that it may last them through the summer. For this purpose they dig a square pit capable of containing seven or eight hundred carcasses. As soon as the ice in the river is of sufficient thickness, it is cut with saws into square blocks, of a uniform size, with which the floor of the pit is regularly paved. The blocks are then cemented together by pouring water in between them, and allowing it to freeze into a solid mass. In like manner the walls are built up to the surface of the ground. The head and feet being cut off, each carcass, without being skinned, is divided into quarters; and these are piled in layers in the pit, till it is filled up, when the whole is covered with a thick coating of straw, which is again protected from the sun and rain by a shed. In this manner the meat is preserved in good condition through the whole summer, and is considered more tender and better flavoured than when freshly killed.

Even in the winter the buffalo continues to range over the plains in a far northern latitude. Mr. Kane mentions seeing a band, numbering nearly ten thousand, at the very northern confines of the Fertile Belt, where the snow was very deep at the time. They, however, had never before appeared in such vast numbers near the Company's establishments. Some, on on that occcasion, were shot within the gates of Fort Edmonton. They had killed with their horns twenty or thirty horses, in their attempt to drive them from the patches of grass which the horses had laid bare with their hoofs. They were probably migrating northward, to escape the human migrations so rapidly filling up the southern and western regions which were formerly their pasture-grounds.

The Cree Indians use dogs to draw their sleighs. They are powerful, savage animals, having a good deal of the wolf about them. They are considered as valuable as horses, as everything is drawn over the snow by them. When buffaloes have been killed in winter, the dead animals are drawn m by them to the camp; and two can thus easily drag a large cow buffalo over the snow. The sleigh or cariole used in these regions is formed of a thin flat board about eighteen inches wide, bent up in front, with a straight back behind to lean against. The sides are made of fresh buffalo hide, with the hair completely scraped off, and which, lapping over, entirely covers the front part, so that a person slips into it as into a tin bath. Each carries but one passenger. The driver, on snow-shoes, runs behind to guide the dogs. Each sleigh is drawn by four dogs, their backs gaudily decorated with saddle-cloths of various colours, fringed, and embroidered in the most fantastic manner, and with innumerable small bells and feathers. Two men run before on snow-shoes to beat a track, which the dogs instinctively follow. A long cavalcade of this description has a very picturesque appearance.

While thus travelling, our friend Mr. Kane caught sight of a herd of buffaloes, which did not perceive the approach of the party till the foremost sleigh was so near as to excite the dogs, who rushed furiously after them, notwithstanding all the efforts of the drivers to keep them back. The spirit of the hunt was at once communicated through the whole line, and the entire party were in an instant dashing along at a furious rate after the buffaloes. The frightened animals made a bold dash at length through a deep snow-bank, and attempted to scramble up the steep side of the river, the top of which the foremost one had nearly reached, when, slipping, he rolled down and knocked over those behind, one on the top of the other, into the deep snow-drift, from which men and dogs were struggling in vain to extricate themselves. It would be impossible to describe the wild scene of uproar that followed.   One of the sleighs was smashed, and a man nearly killed; but at length the party succeeded in getting clear, and repairing the damage.

AN INDIAN STRATAGEM.

In some districts, where the buffaloes can with difficulty be approached, the Indians employ a stratagem to get them within reach of their arrows or rifles.   One of the Indians

They then crawl on all fours within sight of the buffaloes, and as soon as they have engaged their attention, the pretended wolf jumps on the pretended calf, which bellows in imitation of the real one. The buffaloes are easily deceived in this way, as the bellowing is generally perfect, and the herd rush on to the protection of their supposed young, with such impetuosity that they do not perceive the cheat till they are quite close enough to be shot.

On one occasion Mr. Kane and his Indian companion fell in with a solitary bull and cow. On this they made a "calf," as the ruse is called. The cow attempted to spring towards them, but the bull, seeming to understand the trick, tried to stop her by running between them. The cow now dodged and got round him, and ran within ten or fifteen yards of the hunters, with the bull close at her heels, when both men fired, and brought her down. The bull instantly stopped short, and, bending over her, tried to help her up with his nose—evincing the most persevering affection for her; nor could they get rid of him, so as to cut up the cow, without shooting him also, although at that time of the year bull flesh is not valued as food when the female can be obtained. This, and another example which has been given, show that these animals are capable of great affection for each other.

The Indians also occasionally approach a herd from leeward, crawling along the ground so as to look like huge snakes winding their way amid the snow or grass, and can thus get sufficiently near to shoot these usually wary animals.

CHAPTER VI.

RODENTS.

THE BEAVER.

OF all mammals, the beaver is the most especially fitted to enjoy a social life. When in captivity and away from its kind, it appears to possess but a small amount of intelligence; it forms no attachments to its human companions, and is utterly indifferent to all around it. But in its native wilds, associated with others of its race, what wondrous engineering skill it exhibits, and how curious are its domestic arrangements!

It is essentially a hard worker. Other animals sport and play and amuse themselves. What young beavers may do inside their lodges, it is difficult to say; but the elders, from morn till night, and all night long, labour at their various occupations, evidently feeling that they were born to toil, and willingly accomplishing their destiny.

The beaver has fitly been selected as the representative animal of Canada, on account of its industry, perseverance, and hardihood, and the resolute way in which it overcomes difficulties. Certain conditions of country are necessary to its existence, and when it does not find these ready formed, produce them by its own exertions. Where it can find rivers, brooks, and swampy lakes which maintain an even level throughout the year, the beaver has a tolerably idle life; but as in most districts the levels of rivers and lakes are apt to sink at various seasons if left to themselves,—whenever an emigrant party of beavers have fixed on a new locality, they set to work to dam up the stream or outlet of the lake, to prevent a catastrophe which might bring ruin and destruction on their new colony. In Nova Scotia, as well as in other parts of North America, large level spaces are found covered with a rich alluvial soil, from which spring up waving fields of wild grass. From this the human settler draws an abundant supply of hay for his stock in winter, and ought to feel deeply indebted to the persevering beaver for the boon. They are known as "wild meadows," and are of frequent occurrence in the backwoods. It is evident that they were formed by the following process :—They are found in valleys through which, in ages past, a brook trickled. A party of beavers arriving, and finding an abundance of food on the side of the hills, would set to work to form a dam of sufficient strength to keep back the stream, till a pond was created, on the edge of which they might build their dome-shaped habitations. Extensive spaces in the woods were thus inundated, and the colony of beavers lived for long years on the banks of their artificial lakes. They, however, lacking forethought, like many human beings, did not sufficiently look to the future. In process of time the trees, being destroyed, decayed and fell; while the soil, washed down from the surrounding hills, filled up the pond constructed by the industrious animals, and they were compelled to migrate to some other

BEAVERS AND THEIR LODGES.

region, or were destroyed. The dam being thus left unre-
paired, the water drained through it, and the level space was
converted into the rich meadow which has been described.

Beavers' houses, however, are seen in all directions, some-

times on the banks of these artificial ponds, at others by the
sides of large lakes or rivers. Though varying in size, they
all greatly resemble a huge bird's-nest turned upside down.
Some are eight feet in diameter, and three feet in height;
while others are very much larger, being no less than sixteen
to twenty feet in diameter, and nearly eight feet in height on
the outside, and perfectly circular and dome-shaped. The
walls and roofs of these lodges, as they are called, are several
feet in thickness, so that the measurement of the interior
chamber is little more than half that of the exterior. Several
beavers inhabit a large lodge. Their beds, which are separated one from the other, are arranged round the walls, a
space in the centre being left free. The exterior also presents
a very rough appearance, consisting of sticks apparently
thrown loosely together, and entirely denuded of their bark,
as also of branches of trees and bushes closely interwoven and
mixed with stones, gravel, or mud. They are close to the
banks, almost overlapping the water, into which the front
part is immersed. The bottom of the stream or lake is
invariably deepened in the channel approaching the entrance,
thus ensuring a free passage below the ice into the structure.
The tunnel is from two to three feet long. In the inner part
of the hut the materials are laid with greater care, and more
firmly bound together—with mud and grass—than on the outer.
Even in one of the larger houses the chamber—for there is but
one—is only between two and three feet in height, though as
much as nine feet in diameter. It slopes gently upwards from
the water. Inside there are two levels: the lower one may
be called the hall. On this the animals shake themselves
when they emerge from the subaqueous tunnel; and when dry,
clamber up to the upper story, which consists of an elevated

bed of boughs running round the back of the chamber. It is thickly covered with dry grass and thin shavings of wood. The whole of the interior is smooth, the ends of the timbers and brushwood which project inwards being evenly gnawed off. There are always two entrances—the one serving for summer, and letting in the light; while another sinks down at a deeper angle, to enable the owners during winter to get below the water. Beavers are especially clean animals, and allow no rubbish to remain in their abode ; and as soon as they have nibbled off the bark from the sticks, they carry them outside, and place them on the roof of their hut, to increase its thickness, or let them float down the stream.

During the summer they are employed all day in ranging the banks and cutting provisions for their winter consumption, all their architectural occupations being carried on at night. Their winter stock of food consists of short lengths of willow and poplar,—the bark of which only, however, they eat. These they sink with mud or stones in some quiet pool near their lodge, and when required for food they dive down below the ice and bring up as many as are required for family consumption. Besides their lodge, they form in the neigh-bourhood a long burrow sufficiently broad to enable them to turn with ease. The entrance is at a considerable depth below the surface of the water, and extends from ten to twenty feet into the bank. This burrow serves as a safe retreat, should their house be broken into, and thither they immediately fly when their permanent abode is attacked. In summer they regale themselves on the roots of the yellow lilies, as well as on other succulent vegetation, and any fruits the country affords.

But it is time that we should get a look at the curious

animal itself. We may paddle gently in a birch-bark canoe over a calm lake, and conceal ourselves among the tall grass in some quiet cove where the yellow water-lilies float on the tranquil surface. Through the still air of evening, the sound of the distant waterfall reaches our ears. Wood ducks fly by in vast numbers ; the rich glow of the evening sky, still suffused with the gorgeous hues of the setting sun, is reflected on the mirror-like expanse of water. Watching with eager eyes, we see at length the water breaking some forty yards away, and the head and back of an animal appears in sight. Now another, and then a third, come into view. After cautiously glancing around, the creatures dive, with a roll like that of a porpoise, but shortly appear again. Our Indian, pushing the light canoe from amid the grass, paddles forward with eager strokes. One of our party fires, and misses, the echoes resounding from the wood-covered shores, and from island to island, till lost in the distance ; but the cautious animals, forewarned, take good care not to appear again during that evening. We find that our only prospect of examining them is by trapping one in the usual Indian fashion, which we will by-and-by describe.

Mr. Beaver, as the Indians are fond of calling the animal, has a body about three feet long, exclusive of the tail, which is a foot more. He wears on his back a coat of long shining hair, generally of a light chestnut colour, but sometimes of a much darker hue, occasionally perfectly black. Below the hair, next the skin, is a fine, soft, grayish-brown wool. He may be known at once by his broad horizontal flattened tail, which is nearly of an oval form, but rises into a slight convexity on its upper surface, and is covered with scales. His fore-feet are armed with nails, and serve for the purpose of hands—indeed, he vies with the monkey in the use he can make of them. The hind-feet are webbed, and with these—together with his tail, which acts as a rudder—he is enabled to swim rapidly through the water. The beaver is a rodent, with a short head and broad blunt snout, and his incisor teeth are remarkably large and hard, enabling him to bite through wood with wonderful

THE BEAVER.

case and rapidity. So great is their hardness, that formerly the Indians were accustomed to use them as knives for cutting bone and fashioning their horn-tipped spears.

The beaver, it has been said, always chooses banks by the side of a lake or river of sufficient depth to escape being frozen to the bottom, even during the hardest frost. Thus, he can at all times obtain a supply of water, on which his existence depends ; indeed, the bark on which he lives requires to be moistened before it becomes fit for food. When instinct teaches a colony of beavers that the water is not of sufficient depth to escape freezing throughout, they provide against the evil by making such a dam as has been mentioned, across the stream, or the outlet of the lake, at a convenient distance from their habitations. The plan of these dams varies according to the character of the lake or stream. If the current is but slight, they build the dam almost straight ; but where the water runs at a rapid rate, it is almost always constructed with a considerable curve, the convex side towards the stream. Frequently, in such cases, if there is any small island in the centre, it is taken advantage of, and the dam is built out to it from either bank. They make use of a variety of materials ; employing drift-wood when it can be obtained, to save themselves the trouble of cutting down trees. This they tow to the spot, and sink it horizontally with mud and stones. They also employ pieces of green willows, birch, and poplars, intermixing the whole with mud and gravel, in a manner which contributes greatly to the strength of the dam. They observe, however, no order or method in the work, placing their materials as they can obtain them, except that they make the dam maintain its regular sweep, and form all parts of equal strength. They carry the mud and stones in their fore-paws ; and in one night will collect as much as amounts to many thousands of their little loads. When drift-wood is not to be found, they obtain the timber they require from the groves skirting the lake or pond. To do this, they squat on their hams, and rapidly gnaw through the stems of trees from six to twelve or fourteen inches in diameter, with their powerful incisors. Sometimes a tree will not fall prostrate, the boughs being caught by its neighbours. But the beaver is not to be disappointed; he sets to work and gnaws away a little above the first place, thus giving it a fresh start, in order that the impetus may disengage it from the branches which keep it up. The tree being cut up, the beavers, uniting, tow the pieces down to the dam. They then plunge into the water and bring up the mud and small stones with which to keep it sunk. A long constructed dam, by being frequently repaired with fresh mud, becomes at length a solid bank, capable of resisting a heavy rush, either of water or ice; and as the willow, poplar, and birch generally take root and shoot up they by degrees form a regularly planted hedge, which in some places becomes so tall that birds have been known to build their nests among the branches. These beaver

THE BEAVER AT WORK.

dams also form bridges, over which two or three men may pass abreast, and lead their horses, without risk of breaking through. So rapidly do the members of the industrious community labour, that even the most serious damage to their dams, or habitations, is quickly repaired. They always carry the mud and stones in their fore-paws, pressed against their chins, but they drag the wood with their teeth.

The creature does not employ its broad tail, as was once supposed, to plaster down its mud-work, nor does it use it as a vehicle for transporting materials; its sole object being to guide it when in the water, and as a counterpoise, by moving it in an upward direction, to the tendency it would otherwise have of sinking head-foremost. The creatures cover the outside of their houses every autumn with fresh mud as soon as the frost becomes severe. By this means it freezes as hard as stone, and prevents their common enemy, the wolverene, disturbing them during the winter. From the beaver being seen to flap its tail when moving over its work, but especially when about to plunge into the water, has arisen the idea that it uses this member as a trowel. This custom it preserves even when it becomes tame and domesticated, particularly when suddenly startled.

The beaver, says Captain Hardy, travels a long distance from his house in search of materials, both for building and food. He mentions having seen the stumps of some trees which had been felled, at least three-quarters of a mile from the beaver lodges. Its towing power in the water, and that of traction on dry land, is astonishing. The following account shows the coolness and enterprise of the animals, described by a witness to the fact :—The narrator having constructed a raft for the purpose of poling round the edge of the lake to get at the houses of the beaver, which were built in a swampy savannah, otherwise inaccessible, it had been left in the evening moored at the edge of the lake, close to the camp, and about a quarter of a mile from the nearest beaver's house, the poles lying on it. Next morning, on going down to the raft, the poles were missing ; so, cutting fresh ones, he started with the Indians towards the beaver village.   On reaching their abodes, one of the poles was found deposited on the top of the houses.

In a community of beavers there are frequently some who appear to do no work, and are called by the Canadian trappers *Les paresseux*, or Idlers.   They live apart from the rest, taking up their abodes in long tunnels, which they excavate. Several inhabit the same burrow ; and being males, the idea is that they have been conquered in the combats which take place among the males when seeking their mates, and thus, like monks of old, have retired from the world,—or perhaps it may be only for a period, till they have regained sufficient courage and strength to sally forth, and commence a happier existence with the partner of their choice.   They are far more careless of their safety than the other beavers, and are thus easily caught by the trappers.

The body of the beaver contains a curious odoriferous substance, called by the trappers barkstone, but more scientifically "castor," or "castoreum."   It is contained in two little bags about the size of a hen's egg, and is of a brownish, unctuous consistency.   At one time it was supposed to possess valuable medicinal properties.   It is now, however, chiefly employed by perfumers.   The beavers themselves are strangely attracted by this substance, and when scenting it at a distance will invariably make their way to it.   It is said that the inhabitants of a particular lodge go forth, and having rid themselves of their superabundant castoreum at a little distance, return home ; when the beavers of another lodge, scenting the castoreum, proceed to the same spot, and covering it over with a layer of earth and leaves, deposit their own castoreum

the broken leg off, and went away. It was supposed that he
would not come again ; but two nights afterwards he was
found fast in a trap—in each case tempted by the castoreum.
The stake was always licked, or sucked, clean. The substance
seems to act as a soporific, as the' creatures, after tasting it,
always remain a day without coming out of their houses. So
wary generally are the beavers, that a trapper is always careful not to leave his scent on the spot. To avoid this he frequently cuts down a tree, and walks on its branches towards
the edge of the path, afterwards withdrawing it, and plentifully sprinkling water around."

The Indians and Canadian voyageurs eat the flesh of the
beaver, esteeming it, when roasted with the skin on—the hair
having been singed off—the most dainty of dishes. Early in
this century, when beaver fur was much in demand for the
manufacture of hats, upwards of 126,000 skins were exported
from Quebec alone in one year. The warfare long waged
against the unfortunate rodents now goes on with somewhat
diminished activity. A change of fashion—the substitution
of silk for beaver—has probably saved them from utter exter-
mination. The scientific name of their tribe, *Castor*, was long
a popular term for a hat ; but now that their fur has ceased
to be employed as formerly, the term itself appears to have
gone out of use.

THE MUSK-RAT, OR MUSQUASH.

Voyaging along the margin of a lake, we may see on the
shores numbers of little flattened oval nests composed of reeds
and sedges, while numerous holes in the bank, with quantities
of shells, chiefly of the fresh-water mussel, scattered round, show
the entrance to the habitations of the musquash, or ondatra,

called also the musk-rat. As evening approaches, the creatures may be seen in fine balmy weather gambolling on the surface, swimming rapidly here and there, or now and then diving below, apparently fearless of the passing canoe. The little sedge-built hut of the water-rat is constructed much in the same way as the beaver's larger mansion. The creature itself looks somewhat like the beaver, and some of its habits are also similar. It is rather more than two feet in total length, of which measurement about ten inches is occupied by the tail. The upper part of the body is of a dark brown colour, tinged in parts with a reddish hue, while the lower part is ashy gray. Its tail is flattened, but vertical. Like the beaver, it is furnished with an undercoat of soft downy fur. Its safety has been provided for by its peculiar colour, which is so like that of the muddy bank on which it dwells, that a keen eye can alone detect it. Its hinder feet are webbed, the imprint on the soft mud being very similar to that of a duck. With the exception of the flesh of the water-mussel, its food is vegetable. It is a great depredator in gardens, which it has been known to plunder of carrots, turnips, and maize—the stalks of which it cuts close down to the ground.

It is sought for on account of its fur, which is very valuable. The traps are set close to a tree, and when one of the creatures is caught, its companions will instantly attack it and tear it to pieces. Generally, however, in its struggles to get free, it carries the trap under the surface, and is thus drowned.

Audubon, the naturalist, gives us an interesting description of them :—"They are very lively, playful animals, when in their proper element—the water—and on a calm night, in a sequestered pool, may often be seen crossing and recrossing in every direction, leaving long ripples in the water behind them, while others stand for a few moments on tufts of grass, stones, or logs, and then plunge over, one after the other, into the water. At the same time others are feeding on the grassy bank, dragging off the roots of various kinds of plants, or digging underneath the edge. These animals seem to form a little community of social playful creatures, who only require to be unmolested in order to be happy."

It has been proposed to acclimatize these little rodents in England, under the idea that thus a valuable addition to the bank fauna of sluggish English streams would be obtained.

PRAIRIE-DOGS.

Vast cities, with regularly laid streets, are often met with in extensive level spots on the prairie. The inhabitants are, however, not men, but creatures the size of a guinea-pig—rodents—a species of marmot. In their habit of associating together in communities, they put us in mind of the industrious beaver; but they are idle little fellows, evidently liking play better than work. Their heads are not unlike those of young terrier-pups, and their bodies are of a light brown colour. They have little stumpy tails, which, when excited, they constantly jerk up and twist about in a curious fashion. Their habitations are regular cones raised two or three feet above the ground, with a hole in the apex, which is vertical for the depth of two or three feet, and then descends obliquely into the interior. From the peculiar yelp or short squeaky bark which they give, the hunters call them prairie-dogs.

In each separate community, which consists of many thousand individuals, there is a president dog, who seems to have especial charge of the rest. As a stranger approaches, the

creatures who are out of their houses scamper back as fast as
their legs will carry them, and concealing all but their heads
and tails, utter loud barks at the intruder. This done, the
greater number dive out of sight with a curious somersault,
their little tails whisking in the air. The chief dog, and perhaps two or three other sentinels with him, remain on the

PRAIRIE-DOGS.

tops of their houses barking lustily till the enemy gets within
a few paces of them, when they also disappear, and the town
remains silent and deserted. The traveller who wishes to
observe their habits, by lying concealed and silent for a few
minutes, may see after a time some little fellow pop his head
out of his house, when he gives a few barks. It serves as a
signal to the rest that danger has disappeared, and immediately

the others emerge from their houses and begin to frisk about as usual.

The holes of these curious creatures are shared by two very different species of guests, one of which, at all events, must prove most unwelcome. One of these is a little owl, which may be seen sitting in front of the burrows or flying about near the ground; or, when the sun sinks low, hopping through the town, and picking up the lizards and chameleons which everywhere abound. He can apparently do no harm to the inhabitants, if he fails to benefit them. The other inmates are rattlesnakes, who, regardless of any objections which may be raised by the dogs, take possession of their holes, and when the sun shines lie coiled up at their sides, now and then erecting their treacherous heads and rattling an angry note of warning, should a thoughtless pup by any chance approach too near. The Indians suppose that all three creatures live on the most friendly footing; but as the rattlesnakes when killed have frequently been found with the bodies of the little prairie-dogs in their insides, their object in establishing themselves in the locality seems very evident.

The poor little dog, indeed, leads a life of constant alarm, with numerous enemies ever on the watch to surprise him. Hawks and eagles, hovering high in air, often pounce down and carry off unfortunate members of the community in their powerful talons. The savage cayote, or prairie-wolf, when pressed by hunger during the winter, frequently attacks the dome-shaped habitation of the little animal, and with claws and teeth tears to pieces the walls, plunging his nose into the passage which he has opened, and working his way down till he seizes the trembling little inmate, who in vain retreats to the inmost recesses of his abode.

any other, though he wisely prefers keeping within the house while the icy blasts blow across the plains. The creature is especially tenacious of life, and even when shot through the body will manage to gain his burrow at rapid speed. He does not run into it, but, like the rabbit, he makes a jump in the air, turns what looks like a somersault, and, flourishing his hind-legs and whisking his tail, disappears as if by magic. In an instant afterwards, however, his little sparkling eyes and nose may be seen above the ground; and if no stranger is in sight, he, with the rest of the community, will commence gambolling and frisking about, forgetful of his numerous foes and previous alarm. It is very difficult to obtain a specimen of the prairie-dog, as, even if mortally wounded, he generally tumbles into his hole before being captured. The inhabitants of the plain, however, manage to catch the animal alive by dragging a cask of water to one of their holes which does not communicate with the rest of the village. They then pour the water down the hole, either drowning the creature or com-pelling him to come out. He is very soon reconciled to a state of captivity, and after two days appears on the most intimate terms with his captors. Even when turned loose again the creatures will not leave the neighbourhood of the house, but burrow under the foundation, making themselves quite at home, and fearlessly come out to be fed when summoned by a whistle. They become, indeed, very interesting and pretty little pets.

We shall meet with a similar animal on the pampas of South America, and which has also the companionship of a little owl.

There are several other species of marmot in America One is called the QUEBEC MARMOT, which lives a solitary life, making an almost perpendicular burrow in dry ground at a distance from water.

The beautiful little, often-tamed WOODCHUCK, is another American marmot. It makes a deep burrow in the sides of hills, lining the chamber at the inner end with dry leaves and grass. It may frequently be seen by the traveller running rapidly along the tops of fences, as if to keep company with him—now getting ahead, then stopping and looking back to see if he is coming, and then going on again, till, growing tired of the amusement, it gives a last stare and then scampers back the way it has come.

THE PORCUPINE.

Unattractive as the fretful porcupine appears when considered as a means of satisfying man's hunger, it is hunted throughout North America for the sake of its flesh, which forms an especially dainty dish, not only in the opinion of the Indians, but in that of every European who has partaken of it. The creature dwells in small caverns, either under a pile of boulders, or amid the roots of large trees ; but it also, with its sharp claws, easily climbs up the trunks, and may sometimes be seen reposing on their very summits, where it feeds on the bark of the young branches, or the berries when they become ripe.

The Canadian porcupine is also known as the cawquaw or urson. It is nearly four feet long altogether, the head and body measuring upwards of three feet, while the tail is about three inches in length. It is less completely defended with spines than the porcupines of other countries—part of its

THE PORCUPINE.

)ur, the points being dark.
, it can gallop along at con-
ised, generally escapes to its
scrambles up the trunk at a
s to the porcupine's den, by
lso by the ordure outside the
aths lead from the den to its
o a beech grove, on the mast
in the winter-time, to some
Indian hunter also discovers
he bark ; and should he be
·ger game, he seldom fails to
he creature is hunted by the
·em to take great delight in
formidable weapons of their
·aw them out of their dens
·ven the settlers' dogs exhibit

HUNTING THE PORCUPINE.

the same strong fancy for hunting porcupines, but are not so successful in coming off without injury; indeed, they often issue from the combat covered over with spines sticking in their flesh.

Captain Hardy gives us an anecdote of the extraordinary fancy the Indian dogs have for hunting porcupines. One of these dogs was quite blind; and yet, if the porcupine "treed," the little animal would sit down beneath, occasionally barking to inform his master where lodged the fretful one. Another dog was not to be beaten when once on a porcupine. If the animal was in its den, in he went, and, if possible, would haul it out by the tail; if not strong enough, his master would fasten a handkerchief round his middle, and attach to it a long twisted withe. The dog would go in, and presently, between the two, out would come the porcupine.

By the end of the " fall," the animal becomes loaded with fat, from feeding on the berries found in the " barrens." Its cry is a plaintive, whining sound, not very dissimilar to that of a calf moose. The female produces two at a birth early in the spring. The porcupine can easily be tamed ; and Audubon mentions one which was so entirely domesticated, that it would come voluntarily to its master, and take fruit or vegetables out of his hand, rubbing against him as does an affectionate cat. The same animal, however, showed considerable courage. On one occasion it was attacked by a ferocious mastiff. One morning the dog was seen making a dash at some object in the corner of the fence. This proved to be the tame porcupine, which had escaped from its cage. The dog seemed regardless of all its threats, and probably supposing it to be an animal not more formidable than a cat, sprang at it with open mouth. The porcupine seemed to swell up, in an instant, to nearly double its size ; and as the dog sprang upon it, dealt him such a sidewise blow with the tail, as to cause the mastiff to relinquish his hold instantly, and set up a howl of pain. His mouth and nose were full of quills. He could not close his jaws, but hurried, open-mouthed, off the premises. Although the servants instantly extracted the spines from the mouth of the dog, his head was terribly pierced, and it was several weeks before he recovered. The porcupine, however, suffered severely from the combat ; and as the hot weather came on, showed great signs of distress, and finally died of heat.

The quills of the porcupine are brilliantly stained by the Indians with a variety of colours, and are extensively used by their squaws in ornamenting with fanciful patterns the birch-bark ware which they sell to the white settlers.

(279)

CHAPTER VII.

CARNIVORA.

THE BLACK BEAR.

SEVERAL species of the bear tribe inhabit America; the two most numerous of which are the black bear, or musquaw, and the far-famed ferocious grizzly bear of the Rocky Mountains. The black bear is found generally among the forests and plains of the east, though the grizzly also descends from his mountain fastnesses, and makes his way through the low country to a considerable distance from his usual abode. Although the black bear has not obtained the same character for fierceness as his grizzly relative, he often proves a formidable opponent when attacked by human foes, and is also dreaded on account of his depredations among their flocks and herds. He is, indeed, a monstrous and powerful animal, often reaching six feet in length from the muzzle to the tail—the tail being only about two inches long—while he stands from three to three and a half feet in height at the shoulder. He is covered with a smooth and glossy coat of thick hair, without any wool at the base

He does not always wear a black suit ; sometimes he puts on a brown one. When his coat is perfectly black, he has a cinnamon patch on his muzzle. He varies, too, in shape. Occasionally he is long and low, at others his body is short,—and he has great length of limb. Under ordinary circumstances, he restricts himself to a vegetable diet, but is very fond of a small species of snail which feeds on the prairie grass ; and, like others of his relatives, he is greatly addicted to honey. As his feet are furnished with strong sharp claws, he is able to make his way up the trunks of trees to reach his favourite

THE BLACK BEAR.

food. In this object he displays great perseverance and acuteness. However high up it may be, or in positions most difficult of access, he will manage to reach the combs containing the sweet repast. Should the comb be hidden away in the hollow of some aged tree, with an entrance too small for admitting his huge paw, he sets to work with his teeth, and gnaws away the wood till he has formed a breach of sufficient size to allow him to put it in. He is utterly regardless of the assaults of the tiny inhabitants of the comb ; and scooping out their honey and young together with his fore-paws, devours the whole mass.   He will sometimes, when pressed by hunger, break into the settler's barn and carry off sheep, pigs, and small cattle into the neighbouring woods; and so cun-

THE BLACK BEAR IN THE FARMYARD.

ning is he, that it is not often he is overtaken, or entrapped in the snare laid for his capture.

The Indians of Nova Scotia call him Mooin, which reminds us of Bruin.   The Indians throughout the country pay great respect to the bear, having, like the Esquimau, a high opinion of his intellectual powers, and believing that he is in some way related to them, and possessed of an almost human spirit. Still, they do not scruple to kill him; but as soon as the breath is out of his body, they cut off his head, which they place ceremoniously within a mat decorated with a variety of ornaments. They then blow tobacco-smoke into the nostrils, and the chief hunter, praising his courage, and paying a variety of compliments to his surviving relatives, expresses regret at having been compelled to deprive him of life, and his hope that his own conduct has been altogether satisfactory to Mr. Mooin, and worthy of the renown they have both attained.

The musquaw hibernates, like other bears of northern regions, and is very particular in selecting a dry cave for his long winter's nap. At the "fall," he is especially fat, having lived for some time on the beech-mast, blue-berries, and other fruits which grow in great profusion in the forest. He then weighs 500 pounds, and even 600 pounds. The

IN WINTER QUARTERS.

chief part of the fat lies along the back, and on either side, as in the flitch of the hog. There is no doubt that it is by the absorption of this fat throughout his winter fast of four months that he is enabled to exist—at this time evaporation being at a stand-still. Having at length selected a cavern, or the hollow of a decayed tree, for his lair, he scrapes out all the dead leaves, till the ground is perfectly clean and smooth. It must be deep enough to prevent the snow from drifting into it, and free from any water trickling down from above. He objects especially to a habitation which has been occupied by the porcupine, that animal being far from cleanly in its habits. Perhaps also he has an objection to the quills with which the creature is furnished, from their being likely to produce disagreeable wounds. He forgets, perhaps, that the rubbish he has scraped out will betray his abode to the hunter—which it assuredly does. The Indian, on discovering this indubitable sign of Mooin's abode, takes steps to arouse him and plant a bullet in his head, or to batter out his brains with his axe. Mooin, however, in spite of his usual sagacity, ignorant that his abode may be discovered, perhaps already overcome with a strange desire to sleep, crawls in for his winter's snooze. He is frequently accompanied by a partner, who will add to his warmth and comfort. He there lies down with his fore-paws curled round his head and nose, which he pokes underneath his chest. Here he remains asleep till the warm sun of March or April tempts him to crawl out in search of food to replenish his empty stomach and strengthen his weakened frame. Madam Mooin is generally, at this time, employed in the pleasing office of increasing her family. Her young cubs, when born, are curiously small, helpless little beings, not larger than rats. Generally there are two of them, and they are born about the middle of February. She manages to nourish them without taking any food herself till March or April, when she also, like her better half, sallies forth in search of provender. The young creatures grow but slowly, and do not attain their full size till they are about four years old. Even when about a couple of months old, the

watch, as he returns, for an enemy. He creeps up, accord-
ingly, looking on either side, his caution increasing as he
approaches his prey. The hunter, therefore, to outwit h'
seeks his trail in the direction in which he has retreated, and
conceals himself near it, but at some distance from the carc.
He waits till the sun is setting, when he is almost sure to see
the bear come tripping nimbly along, not yet thinking it
necessary to employ caution. At this moment a rifle-bullet,
placed in his head, deprives him of his intended feast and
his life at the same time.

The black bear possesses wonderful strength—said to be
fully equal to that of ten men. Experiments have been tried,
in which so many persons have attempted to drag off a cask
baited with molasses, or other sweet stuff, secured to a rope,
when the bear has carried it away with perfect ease, in spite
of their united efforts to draw it from him.

The most dangerous time to attack a she-bear is in the
spring, when she is accompanied by her cubs. If she has
time, she will lead them off to a place of safety ; but if not,
she will chase the intruder from her domains—and woe betide
him if he cannot manage to escape her claws ! Bears are
easily taken in traps, baited with small bundles of sticks
smeared with molasses. They are hunted in the " fall," when
they have become fat with the ample supply of blue and
whortle berries or beech-mast on which they have been feeding. To obtain the beech-mast, Bruin will frequently climb a
tree, and sometimes, like the orang-outang of Eastern seas,
will build a rough platform for himself among the upper
branches, where he can lie concealed and munch his food at
leisure. The most certain way to obtain the animal in this
case is to cut down the tree and shoot him as he reaches the

the brain—in all probability he will quickly be torn pieces.

The grizzly frequently attains a length of nine feet, and weighs from 700 to 800 pounds. His head, in proportion to his muzzle, is very large. He has a long, narrow muzzle,

THE GRIZZLY AND BLACK BEARS.

somewhat flattened, with large, powerful, canine teeth. His eyes are small, and deeply sunk in his head. His tail is so short, that it is completely concealed by the surrounding hair. He possesses remarkably long feet, which, in the full-grown animal, are eighteen inches in length; and they are armed with sharp and powerful claws five inches long, and so extremely sharp, that they cut into the flesh like knives. He can also use them separately like fingers, so that he can grasp a dry clod of earth and crumble it to dust as a human being could do with his hand. He can also, with them, dig into the ground; and when the weight of his body is not too great, they enable him to climb trees, although not with the speed of his black brother of the plains. As acorns form a portion of his food, it is said that he will climb a tree and shake the boughs vehemently to make them fall, when he descends and revels on the fruit his ingenuity has thus obtained. The hunter who has to fly for his life may however escape from a bear,—when the monster is filled out with autumn food, and cannot manage to raise his huge body from the ground,—by climbing a tree.

The grizzly varies much in colour. Sometimes his fur is of a duilish brown, freckled over with grizzly hairs; while other specimens are entirely of a steely gray. In all cases, the grizzly hairs give a somewhat white appearance to the surface of the fur. When the animal is young, his fur is of a rich brown, and often very long and thick, and much finer than that of the adult animal. When the creature walks, he swings his body in an odd fashion, rolling his head, at the same time, from side to side, which gives him a remarkably awkward look. Although the grizzly occasionally satisfies himself with vegetable diet, he will also attack and devour any animals he can kill. He does not hesitate to assault the powerful bison; and on overtaking a herd, he will spring without hesitation on the largest bull, and, with the tremendous strokes of his powerful paws, speedily bring it to the ground, when he will without difficulty drag the enormous carcass off to his lair, to devour it at his leisure. All other animals stand in awe of the grizzly; and even the largest pack of hungry wolves will not venture to attack him, nor indeed will they touch his carcass after he has succumbed to the rifle of the hunter.   Horses especially are terror-stricken when they scent or see a grizzly; and not until they have been care- fully trained, will they even allow the skin of one to be placed on their backs.

The grizzly employs his claws both in digging for roots and in burying any large animal he may have killed, to pre- serve the carcass till he requires it for another meal.   An anecdote is given of a hunter who, pursued by one of these monsters, took advantage of this propensity to save his life. His rifle was unloaded.   Of course he had not wounded the bear, or his stratagem would have been in vain.   Throwing himself on the ground, the hunter closed his eyes, and stretch- ing out his limbs, feigned to be dead.   It must have been a fearful moment when he felt the bear lift up his body in his claws to carry him away to the neighbourhood of his lair. The bear having dug a hole, placed him in it, and covered him carefully with leaves, grass, and bushes.   An Indian, or hardy backwoodsman, could alone have existed under such circumstances.   The hunter waited anxiously till he heard loud snores proceeding from the cavern.   Then, slipping up, like Jack the Giant-killer from the castle of the ogre, he scampered off as fast as his legs could carry him.

Mr. Kane—the Canadian artist—mentions meeting a grizzly when in company with an old, experienced half-breed hunter, François by name.   François, however, declined firing, alleging that the risk was greater than the honour to be obtained—his own character for bravery having been long established.   Young hunters might do so for the sake of

Indian chief—round his neck.   Although
vo barrels, and François had his rifle, they
hances to one they would not kill him in
hand-to-hand encounter.   The bear walked
n now and then, but seeming to treat them

ore this, a party of ten Canadian voyageurs,
on in the neighbourhood· of the mountains,
ed round a blazing fire, eating a hearty
hen a large, half-famished bear cautiously
oup from behind a chestnut-tree.   Before
of his presence, he sprang across the fire,
men, who had a well-finished bone in his
aist with his two fore-paws, and ran about
hind-legs with him before he stopped.   The
were so thunderstruck at the unexpected
h a visitor, and his sudden retreat with
n—the man who had been carried off—that
e lost all presence of mind, and, in a state
running to and fro, each expecting in his
ped in a similar manner.   At length Bap-
half-breed hunter, seized his gun, and was
; at the bear, when he was stopped by some
told him that he would inevitably kill their
he position he was then in.   During this
king his grasp of the captive, whom he kept
m, very leisurely began picking the bone
pped.   Once or twice Louisson attempted
only caused the bear to watch him more
making another attempt, the bear again

seized him round the waist, and commenced giving him one of those dreadful embraces which generally end in death. The poor fellow was now in great agony, and gave way to the most pitiful screams. Observing Baptiste with his gun ready, anxiously watching a safe opportunity to fire, he cried out, "Tire! tire! mon cher frère, si tu m'aimes! A la tête! à la tête!" This was enough for Le Blanc, who instantly let fire, and hit the bear over the right temple. He fell; and at the same moment dropped Louisson. He gave him an ugly claw along the face, however, which for some time afterwards spoiled his beauty. After he had fired, Le Blanc darted to his companion's side, and with his *couteau de chasse* quickly finished the sufferings of the man-stealer, and rescued his friend from impending death. On skinning the bear, scarcely any meat was found on his bones, showing that it was in a fit of hungry desperation that he had thus made one of the boldest attempts at kidnapping ever heard of in the legends of ursine courage.

WOLVES.

There are several species of wolves in North America : one, a large, black animal, which inhabits the forests; and another, much smaller, which hunts the bison and deer in vast packs across the prairie, and is called the prairie-wolf. Like the wolf of Europe, the black wolf is a fierce, dangerous creature, and equally cowardly. When driven into the corner of a hut, as has sometimes occurred, or when caught in a trap, he will not attempt to defend himself against any person who may enter to destroy him. Audubon mentions an instance of this. A farmer with whom he was staying having lost a number of his animals by wolves, dug several pitfalls in the

PRAIRIE-WOLVES.

. of his farm. Three large wolves were found
ıg in one of these traps. The farmer, instead
əm from above, boldly descended into the trap,
ə creatures one by one by the hind-legs, severed
ɔn, thus preventing their escaping. He after-
and skinned them at his leisure, their skins
cient value to repay him for the loss of his

The prairie-wolves are considerably smaller than their brethren of the woods. They travel in large packs, a solitary one being seldom seen. Their skins are of no value. The Indians will not waste their powder upon them, and they therefore multiply so greatly, that some parts of the country are completely overrun by them. They are, however, caught by pitfalls covered over with switches baited with meat. They destroy a great number of horses, particularly in the winter season, when the latter get entangled in the snow. In

THE CAYOTE.

this situation, two or three wolves will often fasten on one animal, and speedily, with their long claws, tear it to pieces. The horses, however, often bravely defend themselves; and Mr. Goss mentions finding near the bodies of two of these animals, which had been killed the night before, eight wolves lying dead and maimed around,—some with their brains scattered, and others with their legs or ribs broken.

Let us watch from an ambush the manœuvres of a pack of

tection, and by their neighing express their joy and gratitude at our timely interference.

LYNXES.

Although lynxes are not so numerous in America as wolves, they are equally destructive, and individually more daring—attacking deer and smaller animals when they can take them at a disadvantage. They seldom fly, as wolves do, on the first approach of man. In size, the largest does not exceed the dimensions of an English mastiff. The Canadian lynx is frequently termed the Peeshoo, and sometimes " Le Chat " by the French Canadians. His coat is covered with long hairs of a dark gray hue, besprinkled with black, the extremities of which are white, with dark mottlings here and

THE LYNX.

there on the back. Sometimes the fur is of a ruddy chestnut tinge, and the limbs are darker than the rest of the body—which is about three feet long. The animal possesses powerful limbs, and thick, heavy feet, furnished with strong, white claws. When moving over the ground it leaps in successive bounds, its back being slightly arched, and all its feet pitching at the same time. It also swims well, and can cross rivers and lakes a couple of miles broad. Strong as it is, it appears it is easily killed by a blow on the back with a slight stick. It ranges throughout the greater part of the continent, and is shot or trapped for the sake of its fur, which is of considerable value.

THE WOLVERENE, OR GLUTTON.

The wolverene, or glutton, carries off the palm for cunning from all the other animals. It is also more ferocious and daring for its size than even the huge grizzly, while for voracity it is unsurpassed. In appearance, it is somewhat

THE GLUTTON.

similar to a young bear. It is of a brownish-black colour, with a black muzzle and eyes of a dark hue, the space between them being of a brown tint. The paws are also quite black, contrasting with the ivory whiteness of the claws. It possesses large and expanded paws, to enable it to

pass over frozen snow ; indeed, so large are they, that i ...
footsteps are often mistaken for the tracks of the bear.   In
one of its habits it resembles Mr. Bruin, having the custom,
when it finds an animal which it cannot devour at one meal,
of carrying off the remainder and hiding it in some secure
place.

The glutton moves at a somewhat slow pace, and appears
rather deficient in agility ; but at the same time he is persevering and determined, and will range over a wide extent
of country in search of weak or dying animals, stealing unawares upon hares and birds, &c.   When he takes a fancy to
some larger quadruped as it lies asleep, he springs upon it,
tearing open the neck and throat.   He is supposed to prefer
putrid flesh, and the odour which proceeds from him would
lead us to suppose that such is the case.   The trappers look
upon him with especial hatred, as, with his usual cunning, he
seeks out their hoards of provisions in *cache*, and destroys
their marten-traps.   He himself is so sly that he is seldom
caught in a snare.   When he finds one, he approaches it from
behind, and pulling it to pieces from the outside, carries off
the bait.   The marten-hunter will go forth and set a line of
traps, extending to upwards of forty miles in length or circumference.   The wolverene, observing what he is about,
follows at a distance, carefully pulling the traps to pieces as
he leaves them behind, and eating off the heads of the par-
tridges or other birds which have been used as bait, declining
all the time to run his nose into danger.   When a sable or
marten is entrapped, he tears out the dead animal and carries
it away.   It is even supposed that he will attack a hybernat-
ing bear in his den, and manage to kill him before Bruin
has aroused himself sufficiently for his defence.

shoulder. The well-trained dog, however, quickly finds out when roving about the woods at night.

BROUGHT TO BAY.

Let us accompany Audubon on a 'coon hunt. Our native companions have gone before with the dogs, who are baying at the raccoon in an open part of the forest. On our coming up, a singular scene presents itself to us. The flare of our torch seems to distress him. His coat is ruffled, and his rounded tail seems thrice its ordinary size. His eyes shine like emeralds. With foaming jaws he watches the dogs, ready to seize by the snout each who comes within reach. His guttural growlings, instead of intimidating his assailants, excite them the more. He seizes one, however, by the lip. It is a danger-

sanguinary and savage disposition, and commits great havoc
among domestic as well as wild birds, always destroying far
more than he requires ; merely eating off their heads, or lapping up the blood which flows from their wounds.   He commits occasionally ravages in sugar-cane or Indian-corn plantations ; and, climbing with ease, catches birds, and devours
their eggs.   He resembles the squirrel in his movements ;
and, like that animal, when eating, sits on his hind-legs, and
uses his fore-feet to carry his food to his mouth.   A story
is told of a young tame raccoon let loose in a poultry-yard,
when, his natural disposition overcoming his civilized manners,
he sprang on a cock strutting in a dignified fashion among the
hens, and fixed himself on its back.   The bird, surprised at so
unusual an attack, began scampering round the yard, the hens
scattering far and wide in the utmost confusion.   Still the
little animal kept his seat, till he managed to get hold of the
unfortunate cock's head in his jaws, and before the bird
could be rescued, had crunched it up—still keeping his seat, in
spite of the dying struggles of his victim ; and probably, had
he not been bagged, would have treated all the feathered inhabitants of the yard in the same fashion.   When out hunting on his own account, he often hides himself among the
long reeds on the bank of a lake or stream, and pouncing out
on the wild ducks as they swim incautiously by, treats them
as he does the domestic fowls on shore.

He partakes considerably of the cunning of the fox, yet,
like that animal, is frequently outwitted.   A raccoon after
a long chase managed to reach a tree, which he quickly
climbed, with the aid of his claws, snugly ensconcing himself
in the deserted nest of a crow.   In vain the hunters sought
for him, till his long, annulated tail, which he had forgotten

to coil up within the nest, was seen pendent below it; and the poor raccoon was quickly brought to the ground by a rifle ball.

He has gained the name of the lotor,· or the washer, in

RACCOON AND WILD DUCKS.

consequence of his habit of plunging his dry food into water before eating it. He also drinks a large quantity of water. When moistening his food, he grasps it with both his fore-paws, moving it violently backwards and forwards, as a person does washing clothes in a stream. The German naturalists call him the washing-bear. Though savage and bloodthirsty in his wild state, he is frequently tamed; but he is somewhat capricious in temper, and not easily reconciled when offended. It is curious that he should, when domesticated, change his usual custom of sleeping in the daytime and wandering about at night; but this he does, remaining quiet all night, and making his appearance among the inmates of the house as soon as the sun sheds its light abroad. Though in his wild state a fit member for a temperance society, he will when in captivity, as if to recompense himself for his hard lot, drink fermented liquors of all sorts—the stronger and sweeter the better. An old writer on American animals says, in reference to this propensity, that if taken young it is easily made tame, but "is the drunkenest creature alive, if he can get any liquor that is sweet and strong." The same writer states that the cunning raccoon often catches crabs by inserting one of his feet into their holes, and dragging them out as soon as they seize hold of it.

THE AGOUARA, OR CRAB-EATING RACCOON.

In the Southern States we find another species of raccoon, somewhat larger than the former, who is addicted to eating molluscs and crustaceans, whether marine or terrestrial. It is said, also, that when other means fail of obtaining food, he seats himself on a branch hanging low down over some quiet pool, and using his flexible tail as a fishing-line, waits patiently till its end is caught hold of by a snapping turtle or other inhabitant of the water, when, whisking it up, he tears open the creature's shell and devours the luscious flesh with aldermanic relish. The fur is generally of a blackish-gray hue, washed with a tinge of yellow. A blacker tint prevails on the head, neck, and along the spine. His tail, in proportion to the size of his body, is shorter than that of the common raccoon, and is marked with six black rings, upon a blackish-yellow ground.

THE ERMINE.

When we see the judge seated in his richly trimmed robe of ermine —emblem of purity— or call to mind the regal robes of a proud monarch, we are apt to forget that the fur which we so much admire is but that of the detested stoat, turned white during his abode amid the winter's snow of a northern clime. He is not unlike the weasel, especially when clothed in his darker summer dress, but with a less ruddy hue. The edges of the ears and the toes always remain white.

THE AGOUARA FISHING.

He is considerably larger than the weasel, measuring upwards of fourteen inches, including the tail—which is about

four inches long, the tip almost black. He is a bold hunter,
and follows and destroys the hare, and other animals of equal
size. It is said, even, that several together will venture to
attack a man. They are caught in America by traps, which,
giving the animal a sudden blow, kill it without injuring the
skin.

The winter coat of the ermine is produced by the whiten-

ERMINES.

ing of the fur, and not, as was once supposed, by the substitution of white for dark hairs. Probably one cause of this
change of hue may be that the energies of the creature con-
centrate themselves on the vital organs, to enable it to resist
the extreme low temperature of the icy regions it inhabits,
and cannot thus spare a sufficient amount of blood for the

formation of the colouring matter which tinges the hair. Human beings as well as animals become weaker as they increase in age ; and it has been observed that their hair also loses its colour, in consequence of such energies as they possess being required to assist the more important functions of nature. This corroborates the correctness of the former remark.

The ermine, like other species of its genus, has the faculty of ejecting a fluid of a strong musky odour. It is abundant, not only in the barren grounds of the Hudson Bay territories, but is also found in Norway and Siberia.

When the fur is used for robes, or similar purposes, the black tuft at the end of the tail is sewn on at regular distances to the skin, giving to the ermine fur the appearance we are all familiar with.

THE PINE-MARTEN.

The pine-marten, a species of weasel, obtains its name from being found amid pine forests, and from its habit of climbing the trunks of pines in search of prey. It is a fierce and savage creature, choosing to live alone, away from the haunts of man. It is from eighteen to twenty inches in length—with a tail measuring about ten inches—and is covered with long bushy hair. Moving without difficulty among the branches, it seizes many an unfortunate bird in its deadly gripe before its victim can take to flight—robbing also the nest of the eggs within it.

It is common in Europe, as well as in America ; but in the cold regions of the Hudson Bay Company it is hunted for the sake of its skin, which is, when blanched during the winter's cold, scarcely inferior to that of the celebrated sable.

When pursued and overtaken, it stands at bay, exhibiting

its teeth, erecting its hair, arching its back, and hissing like a cat. It forms its burrows in the ground, the female producing a litter of from four to seven. Like other animals of its tribe, it emits a peculiar musky smell.

THE OTTER.

In winter, along the steep banks of the frozen streams, smooth and shining tracks may be readily detected. They are produced by otters, which have a curious habit of sliding downwards for their amusement—much as human beings are accustomed to do in Canada in their toboggings. To do this, they lie on their bellies, with their fore-legs bent backwards, and giving themselves an impetus with their hind-legs, down they glide, at a swift rate, upon the ice. This sport they will continue for some time, climbing up again to the top of the bank, and repeating the process over and over again. They are also accustomed to pass through the woods from lake to lake, making a direct track in the snow. These tracks are easily known. Then comes a broad trail, as if made by a cart-wheel. This is formed by the animal throwing itself on its belly, and thus sliding along over the surface for several yards. These places are called "otter rubs."

There are two species of otter in North America—one on the east, and the other on the Pacific slope—differing slightly from each other. The former is considerably larger than that of the Old World, measuring, from the nose to the tip of the tail, sometimes from four and a half to five feet. Like most other water animals, it possesses two sorts of hair: the one is long and shining, and of a rich brown colour, except on the throat, which is of a dusky white; the other is very fine and soft, lying next the skin, and serving to protect it from the extremes of heat and cold. It has excessively sharp, short teeth, which enable it to hold fast the fish, on which it chiefly feeds. Its body is elongated and much flattened, and the tail, which is of great length, is also flat and broad. The legs are short and strong, and so loosely jointed that it can turn them in any direction when swimming.

The habitations of otters are formed in the banks of rivers

THE OTTER.

or lakes, and are not altogether of an artificial character, as they prefer occupying any deserted hollow or natural crevice to the trouble of digging burrows for themselves. Though they are very playful animals, and delight apparently in sport, they are somewhat of a savage disposition, and must be taken very young to be domesticated. They are cautious, timid animals, and can seldom be approached unawares. They eat all sorts of fresh-water fish, such as trout, perch, eels, and suckers; and will also devour frogs. Occasionally they may be observed on a rocky islet of some lone stream, resting after a banquet, or about to plunge into the water in chase of one of the finny tribe, which their keen eyes detect swimming by.

They are trapped, in Canada, by steel traps, which are submerged close to the bank below their "rubs." They make a peculiar whistling sound, which the Indian can imitate perfectly, and thus frequently induces them to approach. Their

OTTERS FISHING.

skins are manufactured into muffs and trimmings and caps, such as are usually worn in winter by Canadians.

An otter, when attacked, will defend itself with desperation, snapping furiously at the Indian, and then shaking its head violently as a dog does when destroying a rat. Their bite is severe—sufficient indeed to snap off a man's finger—and when once its jaws are closed, no power is capable of making it relinquish its grasp. The Canadians do not attempt to tame the otter; but the persevering Chinese not only contrive to domesticate the species found in their country, but teach them to capture fish for their benefit.

THE SKUNK.

Rambling amid the woods, even in the neighbourhood of

THE SKUNK.

settlements, we may occasionally come upon a curious little animal, with a party-coloured coat and bushy tail, and an amiable and gentle appearance. The creature appears to be in no way timid, and will very likely await our approach. As we draw near it, however, it is apt to turn round and erect its bushy tail perpendicularly. Let us beware of what we are about, for, in a moment, the creature may send over us a

12

shower of a substance so horribly odious, that not only may we be blinded and sickened by the effluvium, but our clothes will be made useless, from the difficulty of getting rid of the odour.

The creature is the skunk, and is about the size of a cat. It possesses short round ears, black cheeks, and a white stripe extending from the nose to the back. The upper part of the neck and the whole back are white, divided by a black line. Below, it is black, as are the legs; and it has a full tail of coarse black hair, occasionally tipped with white. Its legs are short, and it does not possess much activity. Its feet are armed with claws, somewhat like those of the badger.

It appears to use this horrid effluvium—which is generated in glands near the tail—as a means of defence. All other animals have a due horror of it. Anything which it touches is tainted: provisions are destroyed; and clothes, though often washed, will retain the smell for many weeks. At one time this substance was used for medicinal purposes. The mode of defence bestowed on the skunk is somewhat similar to that employed by the cuttle-fish, which emits a dark liquor when pursued. Those who have once smelt the horribly fetid odour of the skunk will not easily forget it.

THE PEKAN, OR WOOD-SHOCK.

Still keeping to the lakes and streams, we may often fall in with a creature of curious habits, which, unlike those just described, lives almost entirely among the branches of the trees. In shape it is somewhat like a weasel, and is the largest of the tree martens. It is known as the wood-shock or pekan, and is also called the black cat, and fisher. This last term is inappropriate, as it is not in any way piscivorous.

It is of a dark brown hue, with a line of black shining hair reaching from the neck to the extremity of the tail. The under parts are lighter ; some entirely white. It possesses also a very large, full, and expressive eye.

Though spending its time among the trees, hunting for its prey, it forms a burrow in the ground for its usual habitation. It lives upon squirrels and rabbits, as well as grouse and other birds and their eggs. Not only does it venture to attack the well-armed porcupine, but it kills the animal, and eats it up, quills and all. The difficulty of accomplishing this appears very great, but there are numerous instances in which pekans have been killed, when their bodies were found full of quills, from which they did not appear to have suffered. They eat up, indeed, both the flesh and bones of the porcupine—the latter being so strong that a small bird cannot crack them. Mr. Downs, the naturalist of Nova Scotia, states that he has frequently found porcupine quills in the stomach of the fisher.

The animal is hunted for the sake of its skin, which is of some value—as also for amusement, especially by boys, as the creature is not sufficiently formidable to cause any great danger to them or their dogs. It is about four feet long, including the tail, which measures about eighteen inches.

THE MINK.

Another denizen on the shores of the fresh waters of Canada is the mink, called also the smaller otter, and sometimes known as the water pole-cat. It may be seen swimming about the lakes, preferring generally the still waters in autumn to the more rapidly-flowing currents of spring. It somewhat resembles the otter, and differs in shape slightly from the marten or ferret. Its teeth, however, are more like those of the pole-cat than the otter; while its tail does not possess the muscular power of the latter animal.

Like the otter, it lives upon fish and frogs, but will occasionally make a marauding expedition into poultry-yards. Its general colour is a dark reddish-brown, approaching in some specimens almost to black on the head; while there is a patch of white, varying in size, under the chin. It is trapped by the settlers both in self-defence and on account of its fur, which is of considerable value, and greatly resembles sable—a good skin often fetching four or five dollars.

MARSUPIALS, OR POUCHED ANIMALS :—THE VIRGINAN OR COMMON OPOSSUM.

The opossum, with its prehensile tail, marsupial pouch, and cunning ways, stands alone for its singularity among all the

VIRGINIAN OPOSSUM.

animals of the American continent. Many of the tribe are found in South America; but the Virginian opossum, the size of a full-grown cat, is larger than all its relatives. The head and body measure about twenty-two inches; and the tail, fifteen. It is covered with a light gray hair of wool-like softness, short on the face and body, but long on the legs. The base end of the tail is thick and black, and is covered with small scales. So powerful is this member that the opossum can hold on with it to the bough of a tree, and even when desperately wounded it does not let go. Its face is long and sharpened, the mouth very determined, and armed with numerous sharp teeth. It has thin, naked, round, and blackish ears, edged with a border of white. It has short legs, the feet being armed with claws, and the interior toes of the hind-feet are flat and rounded.

It has the power of emitting a disagreeable odour when chased or alarmed. When pursued, it makes for the nearest tree; and should it discover the approach of a hunter and his dogs when already up a tree, instead of taking to flight, fond, he will climb the tallest stems, and bite them across, so as to bring the heavy ear to the ground. He will also clamber to some higher branch, and hang down, in search of the fruit growing on the boughs incapable of bearing his weight.

HUNTING THE OPOSSUM.

The quality for which he is chiefly noted is his habit of feigning death. Frequently he is brought to the ground, when there he lies, every limb relaxed, evidently as dead as can be. The knowing hunter will, however, keep his glance on the creature. If he withdraws it for a moment, its eyelids will be seen slowly opening; and should he turn his head for even the shortest space, the creature will be on its feet, stealing away through the under-wood. Though so perfectly an adept at " 'possuming," before attempting to practise its usual ruse it will make every effort to escape from its pursuers. When chased alone by a dog, it will content itself by scrambling up a tree, and sitting quietly on a branch, out of reach, looking down on its canine assailant with contempt as it runs barking furiously below it. The opossum is thus said to be " treed ; " and before long, the barking of the dog brings his master to the spot, when the opossum has to fly for its life to the highest branch it can reach. It is easily captured

OPOSSUM AND YOUNG.

by the rudest style of trap, into which it will walk without hesitation. When " feigning 'possum," it will submit to be knocked about, and kicked and cuffed, without giving the slightest sign of life. The flesh of the opossum is white, and considered excellent—especially in the autumn, when, after feeding amply on the fruits, beech-nuts, and wild berries, of which it is especially fond, it is very fat.

The female opossum builds a warm nest of dry leaves and moss, sometimes in the hollow of a rotten tree, or beneath its wide-spreading roots. She has been known occasionally to take possession of a squirrel's nest; and at other times, that of the Florida rat. When her young—generally thirteen to fifteen appearing at a time—are born, they are extremely small—not an inch in length, including the tail—and weighing only four grains. After a couple of weeks or so, she places them in her pouch, when they grow in size and strength, and in about four weeks may be seen with their heads poked out surveying the world, into which they begin to wander at the end of five or six weeks. When first born, they are the most helpless of little creatures, being both deaf and blind.

The larger number of opossums, however, are to be found in South America, where we shall have an opportunity of further examining them.

CHAPTER VIII.

THE FEATHERED TRIBES OF NORTH AMERICA.

THE BALD OR WHITE-HEADED EAGLE.

THE white-headed eagle takes precedence among the feathered tribes of America,—because he stands first in natural order, and has been selected by the people of the United States as their heraldic emblem. Their choice was, by-the-by, objected to by Benjamin Franklin, on the plea "that it is a bird of bad moral character, and does not get his living honestly." There was justice in the remark, for the bald eagle is a determined robber, and a perfect tyrant. He is, however, a magnificent bird, when seen with wings expanded, nearly eight feet from tip to tip—and a body three and a half feet in length—his snowy white head and neck shining in the sun, and his large, hooked, yellow beak open as he espies, afar off, the fish-hawk emerging from the ocean with his struggling prey. Downward he pounces with rapid flight. The fish-hawk sees his enemy approaching, and attempts to escape; but, laden with the fish he has just captured, in spite of the various evolutions he performs, he is soon overtaken by the savage freebooter. With a scream

The plumage of the bald eagle is of a chocolate-brown, inclining to black along the back, while the bill and upper tail-coverts are of the same white hue as the head and neck. He and his mate build their nest in some lofty tree amid a swamp; and repairing it every season, it becomes of great size. Its position is generally known by the offensive odour arising from the number of fish scattered around, which they have let drop after their predatory excursions. The nest is roughly formed of large sticks, moss, roots, and tufts of grass. They commence making fresh additions to their nest early in the year; and the female deposits her eggs in January, and hatches the young by the middle of the following month. Robbers as they are, the white-headed eagles exhibit great parental affection, tending their young as long as they are helpless and unfledged; nor will they forsake them even should the tree in which their nest is built be surrounded by flames. Wilson, the American naturalist, mentions seeing a tree cut down in order to obtain an eagle's nest. The parent birds continued flying clamorously round, and could with difficulty be driven away from the bodies of their fledgelings, killed by the fall of the lofty pine.

Audubon gives us an account of a savage attack he once witnessed made by an eagle and his mate on a swan :—The fierce eagle, having marked the snow-white bird as his prey, summons his companion. As the swan is passing near the dreaded pair, the eagle, in preparation for the chase, starts from his perch on a tall pine, with an awful scream, that to the swan brings more terror than the report of the largest duck-gun. Now is the moment to witness the display of the eagle's power. He glides through the air like a falling star, and comes upon the timorous quarry, which now, in agony and despair, seeks by varied manœuvres to elude the grasp of his cruel talons. Now it mounts, now doubles, and would willingly plunge into the stream, were it not prevented by the eagle, who, knowing that by such a stratagem the swan might escape him, forces it to remain in the air by his attempts to strike it with his talons from beneath. The swan has already become much weakened, and its strength fails at sight of the courage and swiftness of its antagonist. At one moment it seems about to escape, when the ferocious eagle strikes with his talons the under side of its wing, and with an unresisted power forces the bird to fall in a slanting direction upon the nearest shore. Pouncing downwards, the eagle is soon joined by his mate, when they turn the body of the luck'ess swan upwards, and tear it open with their talons.

Along all the coasts of North America, as also at the mouths of the chief rivers, the white-headed eagle is found watching for his prey. An instance is mentioned of one of these savage birds being entrapped, and falling a victim to his voracity. Having pursued a wild duck to a piece of freshly-formed ice, he pitched upon it, and began tearing his prey to pieces, when the mass on which he stood continuing to freeze, his feet became fixed in the ice. Having vainly endeavoured with his powerful wings to rise in the air, he ultimately perished miserably.

THE WILD TURKEY.

The wild turkey, acknowledged to be the finest of game birds, ranges throughout the forests of the more temperate portions of America. It is the parent of the valued inhabitant of our poultry-yards; and in its wild state utters the same curious sounds which it does in captivity. This superb bird measures about four feet in length. Its plumage, banded with black, gleams with a golden brown hue, shot with green, violet, and blue. Its head is somewhat small, and a portion of its neck is covered with a naked warty bluish skin, which hangs in wattles from the base of the bill, forming a long fleshy protuberance, with hairs at the top.

WILD TURKEYS.

The bird in the States, is commonly known as Bubbling Jock, and is called "Oocoocoo" by the Indians. The female builds her nest in some dry, secluded spot, guarding it carefully, and never approaching it by the same path twice in succession. When first her young are hatched, she leads them through the woods, but returns at night to her nest. After a time she takes them to a greater distance, and nestles them in some secluded spot on the ground. At this time they are frequently attacked by the lynxes, who spring upon them, knocking them over with their paws.

HUNTING WILD TURKEYS.

The wild turkey wanders to a great distance from the place of its birth. "About the beginning of October the male birds assemble in flocks," says Audubon, "and move towards the rich bottom-lands of the Ohio and Mississippi. The females advance singly, each with its brood of young, then about two-thirds grown, or in union with other families, forming parties often amounting to seventy or eighty individuals—shunning the old cocks, who, when the young birds have attained this size, will fight with, and often destroy them by repeated blows on the head. When they come upon a river, they betake themselves to the highest eminence, and often remain there a whole day; for the purpose of consultation, it would seem, the males gobbling, calling, and making much ado,—strutting about as if to raise their courage to a pitch befitting the emergency. At length, when all around is quiet, the whole party mount to the tops of the most lofty trees, whence, at a signal—consisting of a single cluck—given by the leader, the flock takes flight for the opposite shore. On reaching it, after crossing a broad stream, they appear totally bewildered, and easily fall a prey to the hunter, who is on the watch for them with his dogs."

THE OCELLATED TURKEY.

A still more magnificent species of turkey than the one just described inhabits Honduras. It may be distinguished from the common turkey by the eye-like marks on the tail and upper wing-coverts. The naked skin of the head and neck, too, is of a delicate violet-blue, covered with numerous pea-looking knobs arranged in a cluster upon the crown. This is of a pale buff-orange, while there is a row of similar marks over the eye, and others scattered about the neck. The wattle hanging from the neck is of a light orange at the tip. The greater wing-coverts are of a rich chestnut, the feet and legs being of a lake colour. It is somewhat smaller than the wild turkey of the States.

THE CANVAS-BACK DUCK.

The celebrated canvas-back duck, allied to the English pochard, makes its appearance among the numerous rivers in the neighbourhood of Chesapeake Bay about the middle of October, as well as in other parts of the Union. It is at that time, however, thin; but soon grows fat, from the abundance of its favourite food. It is from two to three feet across the wings. Its glossy black beak is large. The head and part of the neck are of a rich glossy reddish-chestnut tint, with black breast. The wing-coverts are gray, and the rest of the body white, marked with a number of transverse wavy lines.

It is especially esteemed at table—and those who have eaten it at the hospitable boards of Americans will acknowledge its excellence; though when, on several occasions, some braces of these birds have been sent to England, they have failed to elicit the admiration due to their merits—in consequence, it is said, of not being accompanied by an American cook.

THE SUMMER DUCK.

The most beautiful of the duck tribe which visits the States is the summer or tree duck of Carolina. It bears a strong resemblance in plumage and habits to the celebrated mandarin duck of China. The birds are found perching on the branches of trees overhanging ponds and streams— a habit not usual in the duck tribe—where they may be seen, generally a couple together, the male in his superb garments of green, purple, chestnut, and white, contrasting with the homely plumage of his mate.

THE PINNATED GROUSE, OR PRAIRIE HEN.

On the open " barrens," where a few tufts of stunted brushwood are alone found, the remarkable pinnated grouse may be seen in great numbers running over the ground. Their backs are mottled with black, white, and chestnut-brown ; and the male has two finely ornamented feathers on the neck, streaked with black and brown. It has also a slight crest on the head, of orange colour, hanging over each eye in a semicircular form ; and naked appendages, which hang down from each side of the neck, and can be filled at the will of the bird by air, so that when puffed out they are like two small yellow oranges. As the breeding-season approaches the males appear, uttering strange cries, puffing out these wattles, ruffling their feathers, and erecting their neck-tufts, as if wishing to appear to the greatest advantage before their mates. They occasionally engage in combats with each other, but their encounters are not often of a bloody description.

They form their nests rudely of grass and leaves, under the shelter of a bush or thick tuft of long grass. The hen lays about fifteen eggs of a brownish-white colour.

The most remarkable feature in the history of these birds is the way in which they assemble, as winter approaches, in vast numbers, to obtain protection from the biting force of the north-west winds which sweep over the Missouri country, by huddling close together.

"As evening draws near," says Mr. Webber, who has observed their habits, "they approach the spot they have fixed on, in the usual manner, by short flights, with none of that whirring of wings for which they are noted when suddenly put up; but they make ample amends for their previous silence when they arrive. From the pigeon-roost there is a continuous roar, caused by the restless shifting of the birds, and sounds of impatient struggling, which can be distinctly heard for several miles. The numbers collected are incalculably immense, since the space occupied extends sometimes for a mile in length, with a breadth determined by the character of the ground. The noise begins to subside a few hours after dark. The birds have now arranged themselves for the night, nestled as close as they can be wedged, every bird with his breast turned to the quarter in which the wind may be prevailing. This scene is one of the most curious that can be imagined, especially when we have the moonlight to contrast

with their dark backs.   At this time they may be killed by
cart-loads, as only those in the immediate neighbourhood of
the slain are apparently disturbed.   They rise to the height
of a few feet, with a stupified and aimless fluttering, and
plunge into the snow within a short distance, where they are
easily taken by the hand.   They will, if disturbed when
they first arrive at a resting-place, change it; but after the
heavy snows have fallen, they are not easily driven away by
any degree of persecution.   By melting the snow with the
heat of their bodies, and by trampling it down, they then
form a kind of sheltering-yard, the outside walls of which
defend them against the winds."

They have, besides human foes, numberless enemies among
the foxes, wolves, hawks, and other birds.   The fecundity of
the survivors, however, keeps pace with the many fatalities to
which they are liable.

THE RUFFED GROUSE, OR AMERICAN PHEASANT.

"This elegant species," writes Wilson, "is known throughout North America.   Its favourite places of resort are high
mountains, covered with the balsam-pine and hemlock."   It
prefers the woods—being seldom or never found in open plains.
They are solitary birds; generally being seen in coveys of
four or five, and often singly, or in pairs.

The stranger wandering through the forest is surprised to
hear a peculiar sound, very similar to that produced by striking two full-blown ox-bladders together, but much louder.
It is caused by the ruffed grouse, who, amusing himself by
drumming, is little aware that it will bring the cruel sportsman
towards him.   The bird produces it when standing on an old
prostrate log.   He lowers his wings, erects his expanded tail,

and inflates his whole body something in the manner of the turkey-cock, strutting and wheeling about with great stateliness. After a few manœuvres of this kind he begins to strike with his stiffened wings, in short and quick strokes, which become more and more rapid, till they run into each other. The sound then resembles the rumbling of distant thunder, dying away gradually on the ear.

The hen is an affectionate mother, and takes every means, when a stranger approaches her nest, to lead him away from the spot.

Wilson describes observing a hen-pheasant depart from this usual custom. He came suddenly upon one with a young bird in her company. The mother fluttered before him for a short time, when suddenly darting towards the young one, she seized it in her bill, and flew off along the surface of the ground through the woods, with great steadiness and rapidity, till she was beyond his sight, leaving him in much surprise at the incident. He searched round, but could find no other birds.

Here was a striking instance of something more than " blind instinct "—by the adoption of the most simple and effectual means for the preservation of her solitary young one —in this remarkable deviation from the usual manœuvres of the bird when she has a numerous brood.

The ruffed grouse is of a rich chestnut-brown, mottled with brown and gray ; while on each shoulder are the curious ruffs, or tufts, from which he obtains his name, of a rich velvety black, glossed with green. The skin beneath them is bare ; the tail is gray, barred with blackish-brown.

Another species of grouse, smaller than the two former, inhabits Canada.

PASSENGER-PIGEONS.

Flights of locusts are often seen passing through the air, like vast clouds, obscuring the sky.   The passenger-pigeon of America appears in almost equal numbers.   The accounts of

PASSENGER-PIGEONS.

their vast flights would be incredible, were they not thoroughly well authenticated.

They are beautiful birds; the males being about sixteen inches in length, the females slightly smaller, and usually of less attractive plumage. The head, part of the neck, and chin of the male bird, are of a slaty-blue colour; the lower portions being also of a slate colour, banded with gold, green, and purplish-crimson, changing as the bird moves here and there. Reddish-hazel feathers cover the throat and breast, while the upper tail-coverts and back are of a dark slaty-blue. Their other feathers are black, edged with white; and the lower part of the breast and abdomen are purplish-red and white. The beak is black, and the eyes of a fiery orange hue, with a naked space round them of purplish-red.

Its chief food is the beech-mast; but it also lives on acorns, and grain of all sorts—especially rice. It is calculated that each bird eats half a pint of food in the day; and when we recollect their numbers, we may conceive what an immense amount must be consumed.

The female hatches only one bird at a time, in a nest slightly made of a few twigs, loosely woven into a sort of platform. Upwards of one hundred nests have been found in one tree, with a single egg in each of them; but there are probably two or three broods in the season. In a short time the young become very plump,- and so fat, that they are occasionally melted down for the sake of their fat alone. They choose particular places for roosting—generally amid a grove of the oldest and largest trees in the neighbourhood.

Wilson, Audubon, and other naturalists, give us vivid descriptions of the enormous flights of these birds. Let us watch with Audubon in the neighbourhood of one of their curious roosting-places. We now catch sight of a flight of the birds moving with great steadiness and rapidity, at a height out of gunshot, in several strata deep, and close together. From right to left, far as the eye can reach, the breadth of this vast

procession extends, teeming everywhere, equally crowded. An
hour passes, and they rather increase in numbers and rapidity
of flight.   The leaders of this vast body sometimes vary their

A CLOUD OF PIGEONS.

course, now forming a large band of more than a mile in
diameter; those behind tracing the exact route of their pro-

are knocked down by men with long poles. Some place pots of sulphur under the trees ; others are provided with torches of pine-knots ; and the rest have guns. The birds continue to pour in. The fires are lighted ; and a magnificent, as well as almost terrifying, sight presents itself. The pigeons arrive by thousands, alighting everywhere, one above another, till solid masses, as large as hogsheads, are formed on the branches all around. Here and there the perches give way with a crash, and falling to the ground, destroy hundreds of the birds beneath, forcing down the dense groups with which every stick is loaded. The pigeons continue coming, and it is past midnight before there is any sign of a decrease in their numbers. The ground in all directions is strewed with branches broken by the weight of the birds which have pitched on them. By sunrise, the enormous multitude have taken their departure, while wolves, foxes, and other animals who had assembled to feast on the bodies of the slain, are seen sneaking off.

Audubon describes the flight of one of these almost solid masses of birds pursued by a hawk ; now darting compactly in undulating and angular lines, now descending close to the earth, and with inconceivable velocity mounting perpendicularly, so as to resemble a vast column, and then wheeling and twisting within their continued lines, resembling the coils of a gigantic serpent. Their assemblages greatly surpass in numbers those of the pinnated grouse already described.

HUMMING-BIRDS.

A considerable number of these gem-like members of the feathered tribe make their appearance in summer, even as far north as Canada, and on the sides of the hills rising out of the "Fertile Belt," within sight of Lake Winnipeg,—a region where snow covers the ground for so many months in the year. The most common, as well as the most beautiful, species of these minute birds, is the ruby-throated humming-bird—a name given to it on account of the delicate metallic feathers which glow with ruby lustre on its throat, gleaming in the sunshine like gems of living fire. From the tip of the bill to that of the tail it measures about three and a half inches. The belly is green, and the upper part of the neck, back, and wing-coverts, are of a resplendent and varied green and gold. The breast and lower parts are white, the wings purplish-brown, and the tail partly of the same colour, with the two middle tail-feathers of vivid green.

In the warm climate of the more southern States, the beautiful little ruby-throat is found throughout the winter; and as the summer draws on, the heat in the northern States suiting its delicate constitution, it migrates in large flocks, appearing in the middle States towards the end of April. Tiny as they are, they pass through the air at a rapid rate, and may be seen moving in long undulations, now rising for some distance at an angle of about forty degrees, then falling in a curve. Their long flights are performed at night, it is supposed, as they are found feeding leisurely at all times of the day. Small as they are, from their rapid flight and meteor-like movements they do not fear the largest birds of prey; for even should the lordly eagle venture into their domains, the tiny creatures will attack him without fear : and one has been seen perched on the head of an eagle, at which it was pecking furiously away, scattering the feathers of the huge bird, who flew screaming through the air with alarm, to rid himself of his tiny assailant.

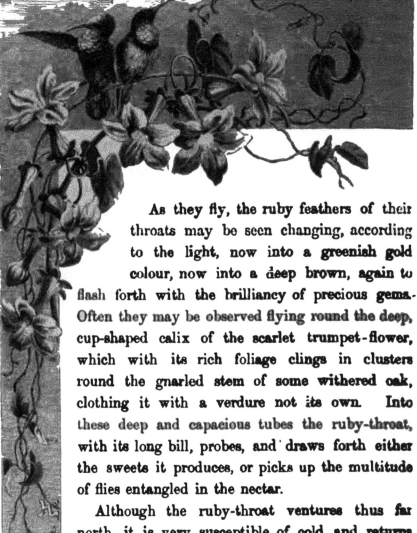

As they fly, the ruby feathers of their throats may be seen changing, according to the light, now into a greenish gold colour, now into a deep brown, again to flash forth with the brilliancy of precious gems. Often they may be observed flying round the deep, cup-shaped calix of the scarlet trumpet-flower, which with its rich foliage clings in clusters round the gnarled stem of some withered oak, clothing it with a verdure not its own. Into these deep and capacious tubes the ruby-throat, with its long bill, probes, and draws forth either the sweets it produces, or picks up the multitude of flies entangled in the nectar.

Although the ruby-throat ventures thus far north, it is very susceptible of cold, and returns southward immediately the summer green of the forest gives place to the golden tints of autumn.

Brave and high-spirited as is the little bird, it is easily tamed ; and Mr. Webber, the naturalist, after many attempts, succeeded in securing several of the species. The first he caught did not flutter, or make the least attempt to escape, but remained quietly in his hand; and he saw, when he opened it, the minute creature lying on his palm, perfectly motionless, feigning most skilfully to be dead; indeed, actually playing "'possum." For some time he watched it with breathless curiosity, when he saw it gradually open its bright little eyes to ascertain whether the way was clear, and then close them slowly as it caught his glance upon it. When a mixture of sugar, water, and honey was brought, and a drop placed on the point of its bill, it came very suddenly to life, and in a moment was on its legs, drinking with eager gusto of the refreshing draught from a silver tea-spoon.

The nest of the ruby-throat is of a most delicate nature; the external parts being formed of a little gray lichen found on the branches of the trees, glued together by the saliva of the bird, and neatly arranged round the whole of the nest, as well as to some distance from the spot where it is attached to the branch or stem itself. The interior is lined with a cottony substance; and the innermost, with the silky fibres obtained from various plants. Within this little nest the female humming-bird lays two white and nearly oval eggs; generally raising two broods in the season. In one week, says Audubon, the young are ready to fly, but are fed by the parents for nearly another week. They receive their food direct from the bill of their parents, who disgorge it in the manner of canaries and pigeons. It is my belief that no sooner are the young able to provide for themselves than they associate with other broods, and perform their migrations apart from the old birds, as I have observed twenty or thirty young humming-birds resort to a group of trumpet-flowers, when not a single old bird was to be seen.

The plumage of the female is in most respects like that of the male, except that she is not possessed of the brilliant feathers on the throat which especially distinguish him.

Although there are several other species of humming-birds which live permanently in the Southern States of the Union, or migrate northward in summer, we cannot now inspect them. We shall have, however, many opportunities of examining a number of the species when we come to visit South America. Although the number of birds and the variety of their species inhabiting North America is very great, except those we have mentioned, they do not in general possess any very interesting peculiarity, which might tempt us to linger longer amongst them, and we will therefore wander on and inspect some of the curious reptiles which inhabit various parts of the American States and Canada.

THE COW-BIRD.

The well-known spring visitor to the woods of England,—the cuckoo,—is undoubtedly destitute of family affection, as are others of its relatives; but this is not the case with the whole tribe. As the spring advances, from the sylvan glades of Pennsylvania a curious note, constantly repeated, is heard, resembling the word "cow-cow." It is the note of a bird, and from the sound it resembles it is generally known as the "cow-bird." It is also called the "yellow-billed cuckoo." It is in no respect behind any of its neighbours of the grove in conjugal and parental affection, for it builds its nest, hatches its own eggs, and rears its own young, Wilson assures us. It is about a foot in length, clothed in a dark drab suit with a silken greenish gloss. A ruddy cinnamon tints the quill-feathers of the wings; and the tail consists partly of black feathers tipped with white, the two outer ones being of the same tint as the back. The under surface is a pure white. It has a long curved bill of a grayish-black above, and yellow beneath. The female differs from the male in having the central tail-feathers of a drab colour, while the under part of her body is of a grayish tinge.

Early in the spring the males frequently engage in desperate battles. After these contests are decided, the couples, pairing, begin building their nests, generally among the horizontal branches of an apple-tree. It is roughly formed of sticks and twigs. On this bed the eggs, three or four in number, of a uniform greenish-blue, are placed. While the female is sitting, the male is generally not far off, and gives the alarm by his notes should any person approach. The female sits so close, that she may almost be reached by the hand, and then suddenly precipitates herself to the ground, feigning lameness—to draw away the intruder from the spot—fluttering her wings, and tumbling over in the manner of a partridge, woodcock, and some other birds. Both parents unite in collecting food for the young. This consists, for the most part, of caterpillars, particularly such as infest apple-trees. They are accused, and with some justice, of sucking the eggs of other birds,—like the crow, blue jay, and other pillagers. They also occasionally eat various kinds of berries ; but from the circumstance of their destroying numbers of very noxious larvæ, they prove themselves the friend of the farmer, and are well deserving of his protection.

THE BLUE-BIRD.

While the robin redbreast cheers us in England during winter with its song, the beautiful little blue-bird performs the same office with its rich sweet notes to the inhabitants of the United States; arriving from Mexico, and still further off regions, as soon as the first signs of approaching spring appear—even before the snow has melted away. Associating fearlessly with human beings, it holds the same place in their affections as the robin.

It is about seven inches long—a rich azure-blue covering the whole upper surface of the head and neck, while the quill-feathers of the wing and tail are jet-black. The throat, breast, and sides are of a ruddy chestnut, the lower portion of the body being white. It builds its nest in the hollow of a decayed tree, sheltered from the rain and cold, and there deposits from four to six eggs at a time, generally rearing two, and sometimes three broods in the season. Its food consists chiefly of spiders and small worms, and soft fruits and seeds.

It is a hardy little bird, and makes its way through all parts of the United States; sometimes, indeed, remaining through the whole winter, when it takes shelter in some warm hollow beneath the snow, from whence, when the sun shines forth, it comes out to enjoy its warmth, and to sing a few cheerful notes. It is especially interesting to watch it take care of its nest and young; perching near them and singing merrily, occasionally flying off to procure a caterpillar for their gaping mouths.

So confiding is the blue-bird, that when a box with a hole in it is arranged in some convenient situation near a house, it will at once take possession, building its nest in it, and never failing to utter its sweet music in acknowledgment of the boon.

THE SNOW-BIRD.

As the cold winter approaches, large flocks of little birds

about six inches in length, with snow-white breasts and slaty-brown or blue backs, make their appearance in the neighbourhood of villages and farm-houses; sometimes, indeed, coming into towns as familiarly as sparrows. Their habits are very like those of sparrows; and when the snow deepens, they mix with them, searching together for the seeds in the sheltered corners of the fields, and along the borders of creeks and fences.

They differ from the snow-bunting of the far north, with which they must not be confounded. In the summer they make their way to the northern regions in large flocks, and build their nests together, being of a very sociable disposition.

THE CAROLINA PARROT.

While viewing the birds of North America, we cannot pass by the well-known, handsome Carolina parrot, which is, notwithstanding its common name, a species of macaw. Large numbers of these beautiful birds are seen winging their way in compact bodies through the Southern States, flying with great rapidity and uttering a loud outrageous scream, not unlike that of the red-headed woodpecker. Sometimes their flight is in a direct line, but generally they perform a variety of elegant and serpentine meanders in their course through the air. Often they may be seen pitching on the large syca-more-trees, in the hollow trunks of which, as also among the branches, they generally roost—frequently forty and more together. Here they cling close to the side of the tree, holding fast by claws and bill. No creatures can be more sociable, and they may be observed scratching each other's heads and necks, and always nestling closely together.

Their plumage is mostly green washed with blue, but the forehead is of a reddish-orange—as are the shoulders, head, and

CAROLINA PARROT.

wings, while the neck and back of the head are
of a bright golden yellow. The wing-coverts
are yellow tinged with green. The bird is
about twenty-one inches long. The female is
much like the male. She makes her nest in
the hollows of trees.

The Carolina parrot exhibits great amiability
of disposition, and is easily tamed, becoming
much attached to those who treat it kindly.
It also exhibits the most extraordinary affection
for its own race. Wilson the naturalist, having
obtained one while on a journey to the Far
West, brought it home upwards of one thousand miles in his pocket. It quickly learned
to know its name, and would immediately come
when called. Procuring a cage, he placed the
parrot under a piazza, where, by its call, it soon attracted the
passing flocks of its relatives. Numerous parties frequently

alighted on the trees immediately above, keeping up a constant conversation with the prisoner. One of these was wounded and captured. Poll evinced the greatest pleasure on meeting with this new companion. She crept close up to it, chattering in a low tone of voice, as if sympathizing in its misfortune, scratching its head and neck with her bi.'—at night, both nestling as closely as possible to each other, sometimes Poll's head being thrust amongst the plumage of the other. The stranger, however, died, and Poll appeared restless and inconsolable for several days. On a looking-glass, however, being procured, the instant she perceived her image all her former fond-

NEST OF THE CAROLINA PARROT.

ness seemed to return, so that she could scarcely absent herself from it for a moment. It was evident she was completely deceived. Often when evening drew on, as also during the day, she laid her head close to that of the image in the glass, and began to dose with great composure and satisfaction.

On another occasion several of these birds were shot down, when the whole flock swept rapidly round their prostrate companions, and settled on a low tree within twenty yards of them. Although many were killed, the rest, instead of flying away, continued looking down at their dead companions with manifest signs of sympathy and concern.

They render the farmer great service, by eating the cockle-burs which grow on the rich alluvial soil of Carolina. This prickly fruit is apt to come off on the wool of the sheep, which, in some places, it almost completely destroys. The bird also lives on the beech-nut and seeds of the cypress. The head—with the brains—and intestines of the Carolina parrot are said to be poisonous to eat ; but how far such is the case seems to be a matter of doubt.

Its chief abode is along the shores of the Mississippi, and it reaches the neighbourhood of Lake Michigan ; but eastward of the Alleghany Mountains it is seldom met with further north than the State of Maryland. Far more hardy than the generality of the parrot tribe, a flock has been seen facing a snow-storm along the banks of the Ohio.

CHAPTER IX.

REPLILES.

TORTOISES :—THE LETTERED TERRAPIN.

TAKING the reptiles in their natural order, we must begin with the tortoises. There is a group of these slow-moving reptiles called terrapins in North America. One of the most common is the lettered terrapin, which inhabits rivers, lakes, and even marshes, where it lives on frogs and worms. It is especially detested by the angler, as it is apt to take hold of his bait, and when he expects to see a fine fish at the end of his line, he finds that a little tortoise has hold of it.

The back is of a dark brown, the edges being ornamented with scarlet marks, like some Eastern alphabet in form.

THE CHICKEN TORTOISE

Large numbers of these little tortoises, about ten inches in length, are seen basking together on the logs or stones on the borders of lakes or streams. The slightest noise arouses them, when they slip off, splashing in all directions into the water. They swim with their little heads above the surface at a rapid rate, bearing a strong resemblance to water-snakes. The creature takes its name from the similarity of its flesh to that of a chicken. It is consequently in great requisition as food.

THE SALT-WATER TERRAPIN.

Another species—the salt-water terrapin—lives in the salt marshes and ponds. It is brown above, and generally yellow below—the lower jaw furnished with a sort of hook. The sides of the head are white, sprinkled with black spots.

THE BOX TORTOISE.

The peculiarity of this creature is that it can draw its head within its shell, so that, as few creatures would wish to swallow such a morsel whole, it has no enemy except man to fear. It might, to be sure, run the risk of being carried off by an eagle and let drop on a hard rock, if the savage king of birds ever does perform such a feat; but though stories are frequently told of his doing so, their truth is greatly doubted.

The box tortoise lives on shore among the pine-forest lands, away from water, to which it seems to have an especial dislike. It is frequently called, therefore, the pine terrapin. It is one of the smallest of its tribe—being little more than six inches long—and varies very greatly in its colour. Its head is remarkable for having a somewhat broad hook at the end of the upper jaw—the lower jaw being slightly hooked.

THE MUD TORTOISE.

The mud tortoise is smaller than the box, being scarcely four inches in length. It can, however, move with considerable speed, and is seen floundering about in the ponds and muddy places, where it searches for aquatic insects, and sometimes even fish, on which it lives. It also vexes the angler by taking hold of his hook, and remaining so quietly sucking in the bait, that only when he hauls˙it up, and the tortoise begins to pull and kick violently about, does he discover his mistake.

It is remarkable for exuding a strong musky odour, from which circumstance it has obtained the name of "stink-pot."

THE ALLIGATOR TERRAPIN.

This giant of its tribe, from the great likeness it bears to the alligator, has appropriately been called after the huge saurian. It has a large head covered with a hard wrinkled skin, and a long thick neck, over which are scattered a number of projecting tubercles. On the shell of the adult animal there is a depression along the centre, which leaves a sort of keel on each side of the central line.

The creature is exceedingly voracious, feeding on fish, reptiles, or any animal substance. It generally inhabits stagnant pools or sluggish streams, living mostly at the bottom. Occasionally, however, it rises to the surface, and elevating the tip of its pointed snout above the water, floats along with the current. Sometimes, indeed, it lands, and makes its way to some distance from the river; but its motions are very awkward, not a little resembling those of the alligator.

A considerable number are taken by strong hooks, and, as the flesh is esteemed for food, are sold in the market.

THE SNAPPING TURTLE.

Although the last-named creature is sometimes called the snapping turtle, the animal to which the name appropriately applies is a very different creature. Its other name is the fierce trionyx.

It belongs to the family of tortoises, popularly called soft turtles. Its flattened head is rather oval, with horny jaws, and hanging fleshy lips, the mouth lengthened into a cylindrical snout. It has an extremely long neck, which it can contract at will; short, wide feet; and toes connected by strong webs. It is the most savage and formidable of its tribe; being terribly destructive, not only among fish, but smaller quadrupeds, birds, and reptiles, which it can capture. For this object it lies in wait till they come down to drink, or till some water-fowl flies too close to its haunt. It is said even to capture and eat young alligators.

Though devouring so many other creatures, the snapping turtle is often eaten himself; being hooked and drawn on shore by the fishermen. It fights, on such occasions, and struggles ferociously, darting its head here and there, endeavouring to seize the hands of its captors with its formidable jaws.

It possesses extraordinary tenacity of life; and even after the head is cut off, the body, it is asserted, will crawl for a short way over the ground.

LIZARDS:—THE SIX-LINED TARAGUINA.

We shall find several lizards in various parts of America—the greater number in the Southern States. The first we meet with is the six-lined taraguina, belonging to the family of teguexins, which are remarkable for the many-sided shields which cover their heads, and the double collar on the throat. This little creature is much smaller than the rest of its family—being only about eleven inches in length—of a darkish green or brown colour, with six narrow yellow streaks along its body, one of which on each side reaches from the eye to the middle of the tail. The lower part is of a silvery white hue, with a bluish tinge in some parts.

It is an excessively lively, active animal, living in dry and sandy places, where it may be found searching for insects. As it is very timid, it takes to flight at the slightest sound, and is not easily caught.

THE GLASS SNAKE.

As the spring comes' on, and the warm sun bursts forth, a formidable snake-like creature, nearly three feet in length, is often seen frequenting the plantations of the sweet potato, or coiled up beneath the roots of an old tree; its keen eye watching for any small reptile or insect which may be passing. The head is small in proportion to the body, and of a pyramidal form—mottled at the sides with black and green, the jaws edged with yellow. Its abdomen is bright yellow; and the upper part of the ear is marked with numerous lines of black, green, and yellow.

Altogether, it has a very venomous look about it; but is truly one of the most harmless of creatures, not being a snake at all, though it goes by the name of the glass snake. It is in reality a lizard; though—not having the vestige of limbs— it is appropriately called the lizard-snake. It has, however, eyelids; and the tongue is not sheathed at the base, as is the case with serpents; while its solid jaw-bones do not enable it to open its mouth, as they are capable of doing. It has a tail twice the length of its body, from which it can with difficulty be distinguished.

Its peculiar characteristic is its extraordinary fragility,—

arising from the muscles being articulated quite through the vertebræ. If struck with a switch, the body is easily broken in two or more parts. Sometimes, indeed, the creature breaks off its own tail, by a remarkable habit it possesses of contracting the muscles with great force. The common English blind-worm breaks to pieces in a similar manner.

THE ANOLIS.

Among the true lizards is a pretty little creature known as the green Carolina anolis. It is especially daring; not only refusing to run away at the approach of man, but will enter

houses, and run about the room in search of flies. It is very active, climbing trees, and leaping from branch to branch in its search for insects, of which it destroys great numbers. It is about seven inches long—mostly of a beautiful green above, with white below; and it has a white

THE ANOLIS.

throat-pouch, which generally appears with a few bars of red upon it, but when inflated the colour spreads over the whole surface.

Mr. Gosse describes one which he saw running about among the branches of a sassafras, just as it had seized a grasshopper. He caught the creature, which was then of a green hue; but, on placing it on an old log, the colour changed to a brownish-black. He was told, that if placed on a green leaf it would again become green. In a short time, after remaining in the sunshine, it changed once more to green. Again it became almost black ; and shutting it up in a desk, after half an hour he was no less surprised than delighted to see the lizard of a brilliant green, the line down the back only being blackish.

When the animal is excited, the pouch, swelling out, becomes of a crimson colour. It is covered with excessively small—scarcely perceptible—scales.

These little creatures are at times very quarrelsome, and will fight together, frequently both the combatants losing their tails in the contest ; while their pouches swell out as they leap at each other and struggle furiously.

THE CROWNED TAPAYAXIN.

This is the scientific name of a creature generally known under the title of the horned-toad, though really a lizard. Its head is of a light brown, marked with dark spots, the under part being of a dull yellow ; and is armed with long conical spines, set round the edge and pointing backwards. The back is covered with shorter and stouter spines, of a triangular shape, extending to the very point of the tail— also armed with a strong row of spines, which gives it a completely toothed appearance. The colour of this curiously-covered back is gray, with irregular bands of chestnut-brown across it.

Formidable as it looks, it is not only harmless, but never retaliates when attacked, and remains perfectly quiet when taken in the hand. It is also easily tamed, and learns to know its owner, and to take food from his hand—preferring little red ants, though it eats readily beetles, flies, and other insects. From its small, rounded form, and the mode of sitting, it has in all likelihood gained its common name of the horned-toad.

SNAKES :—THE RATTLESNAKE.

Throughout North America there are no small number and variety of venomous snakes. The rattlesnakes are perhaps the most numerous, frequenting all parts of the country, though they generally keep to the uninhabited portions. They are found on the northern shores of Lake Superior— though the ground is covered for several months in the year with snow—and often appear in the regions to the west, in the same latitude, up to the Rocky Mountains. They would render some districts uninhabitable, were it not for the signal-giving rattles with which they are armed. Even quadrupeds are alarmed at the sound, and endeavour to make their escape from them ; and horses, it is said, lately arrived from Europe, show the same dread of these deadly serpents as do those born in the country, so that nothing will induce them to pass within striking distance of the creatures.

The wanderer through the forest starts back with dismay as he comes suddenly upon one of these venomous reptiles, and hears its ominous rattle when too near to escape. He must muster all his nerve, and strike it with his stick as it springs ; for a wound from its fangs will, as he knows, bring certain death, far away from human aid.

The rattlesnake,. like others of its tribe in cold regions, hibernates in winter ; and as the autumn comes on, seeks some convenient crevice in which to pass the cold season— generally in the neighbourhood of marshy ground, where it can cover itself up in the masses of a peculiar species of moss growing in such situations. The reptiles are here, during the winter, frequently hunted out and destroyed. At that time, too, their bite is much less dangerous than in the summer—

the amount of venom appearing to decrease with the increase of cold.

THE BANDED AND MILITARY RATTLESNAKES.

Besides the common rattlesnake, there is another known as the banded rattlesnake, and a third species called the small,

ENCOUNTER WITH A RATTLESNAKE.

or military rattlesnake. The latter is more dreaded, from being of less size, and not so easily killed as the former. The sound made by its rattle is extremely feeble, so that it cannot be heard at any great distance. However, as we shall pay more attention to the serpent tribe when we visit South America, where the rattlesnake is also found, we will wait till then to inspect the formation of its rattle, and its other peculiarities.

THE CORN SNAKE.

There are many more harmless than venomous snakes in North America. One of the handsomest of its tribe is the corn snake, belonging to the family of the Colubrinæ. As it avoids the daylight, though very common, it is not often seen in a wild state.

It is, however, frequently tamed by the inhabitants of farm-houses—when it makes itself perfectly at home, and is even of more service than a cat in devouring rats and mice; though occasionally, if a young chicken come in its way, it may gobble it up. This it can easily do, as it is of great size—varying from five to six feet in length. The colours of its body are remarkably brilliant; the general tint being a rich chestnut red, with large patches of a still brighter and deeper red edged with black running along each side, and a second row of smaller spots of golden yellow, alternated with larger ones. The lower portion of the body is silvery white, checkered with black.

THE THUNDER SNAKE.

No fiercer-looking member of the snake family exists in North America—with its mottled head, and black and white body, four feet at least in length—than the quarrelsome thunder snake. From the chain-like markings on its body, it is sometimes called the chain snake; and by others the king snake, on account of its tyrannical disposition.

Though fangless, it is fierce and bold, and has been known

to attack, kill, and eat a rattlesnake; indeed, it will assault any member of its family, if not of its own species, even though but little smaller than itself. It feeds on small quadrupeds, birds, and reptiles; and few human beings who see it moving amid the shady places it inhabits, would fail to get out of its way as quickly as possible.

THE CHICKEN SNAKE.

The bright golden brown chicken snake—marked with narrow stripes along the back, and from four to seven feet in length—in spite of its beautiful and fangless mouth, is an unwelcome visitor in farm-houses when it comes as a stranger, for it is apt to carry off fowls from their roost—as well as their eggs—and will eat up a brood of ducklings without ceremony.

However, as it is of an amiable disposition, it can easily be tamed; and then, having learned good manners, it becomes a favourite, and recompenses its protectors by killing the rats and mice which frequent their premises.

THE MILK OR HOUSE SNAKE.

The beautiful blue house snake—four feet in length, with rows of spots on its side—is often mistaken for the corn snake, its habits being very similar. The lower part of the body is of a silvery white, tesselated with oblong marks of black. The ignorant fancy that it sucks the milk from the udders of the cows, and hence its name; though, probably, it has no objection to a little milk, if it finds it in a pan. Its object, however, in entering houses and farms, is to search for mice and insects, on which it in reality feeds, never interfering with the cows or other animals.

THE BLACK SNAKE.

In many parts of the country, the black snake, on account of its rapid movements, is called the "racer." Though fangless, it often, in consequence of the way in which it rustles its tail among the dry herbage, making a sound similar to that of the rattlesnake, gives no small alarm to the wanderer among the brushwood near the edges of streams or ponds. It is also frequently encountered in the fields or on the roads.

It is generally from five to six feet in length; of a blue-black above, and an ashy gray below. It climbs trees in search of birds or their eggs; and if interrupted in its employment, will turn its rage against the intruder. Sometimes, it is asserted, it will, to his horror, leap down and give him a bite; though the only injury likely to arise is that to his nervous system from fright. Its bite is, indeed, perfectly harmless; and it does good service in hunting rats which live in the outbuildings, being able to climb walls and insinuate itself into the most intricate passages when chasing them.

THE COACH-WHIP SNAKE.

The last snake we will mention is the coach-whip snake, belonging to the family of Dryadidæ. No serpent can surpass it in the rapidity of its movements, as, with its lithe, black body—between five and six feet in length—and whip-like tail, it makes its way amid the grass in pursuit of its prey. It seems literally to fly over the ground with the speed of lightning.

It is curiously like the thong of a whip, being very long in proportion to its girth, with a remarkably small head and neck; its smooth scales—so arranged as greatly to resemble

THE COACH-WHIP SNAKE.

the plaited leather of a whip—of a polished brown-black hue increasing the resemblance.

When about to seize its prey, it darts forward with open mouth, grasping the animal; in an instant it winds its lithe body and tail round and round it, so as to make escape impossible. It will thus attack birds of prey of considerable size, and come off victorious.

Travellers unacquainted with the reptiles which haunt the wilds of America, on first seeing a whip-snake rapidly approaching, will, with sensations of alarm, urge on their steeds to escape—for it appears fully capable of springing up and inflicting mortal injury; but, from having no fangs, it is unable to harm any one. From the delicacy of its colour, the elegance of its form, and the rapidity and gracefulness of its movements, it cannot fail to be admired.

FROGS :—THE BULL FROG.

We shall find no small number of the frog race throughout America. Worthy of being the president of his nation is that enormous batrachian, the bull frog, both from his size, the power of his notes, and his hardihood and endurance. If we visit at night the neighbourhood of some pool or marsh, we shall soon learn to know the sound of his voice, especially when perhaps he and five hundred of his family are, with their heads half out of the water, amusing themselves in the performance of a concert, each striving to outdo his neighbour in the loudness of his tones. He is a first-rate swimmer; and when driven out of the hole in which he passes the warm hours of the day, he plunges into the water, and skims along the surface some distance before he dives below it. Only on such occasions, or when, perhaps, a dark thunder-cloud shrouds the sky, does he appear in the day-time, and give utterance to his notes.

He feeds on snails and water creatures; sometimes on crayfish and other crustaceans; and occasionally, if a duckling or young chicken come in his way, he will not scruple to take them into his capacious maw.

His ordinary size is from six to seven inches; but specimens have been met with which have measured nineteen—and even twenty—inches, from the nose to the extremity of their feet. He has a smooth black skin above, with a greenish hue on the head, and lower part of the body grayish-white—the throat being white, dotted with green. He can take enormous leaps; and is so admirable a swimmer, that specimens have been known to exist in the water without once landing for several years.

THE SOLITARY FROG.

Inland, where no water is to be found, we shall meet with a creature of an olive colour—the back covered with tubercles—and with a blunt nose. It might easily be mistaken for a toad, though it is a veritable frog. Even in winter, before the snow has disappeared, we may see the hardy little creature making its way over the frozen surface of the ground. At the breeding season, however, it returns, like other frogs, to the water. It resides for the chief part of the year in sandy districts, in which it forms burrows, about six inches in depth, by means of a flat, sharp-edged spur, with which it is furnished. Into these burrows it makes its way backwards, very much as a crab crawls into its hole when seeking shelter from danger. There it sits, with its head poked out, watching for passing prey.

THE SAVANNAH CRICKET FROG.

Both in the Northern and Southern States we shall find a merry little creature, with a voice greatly resembling that of the cricket. Living near the borders of stagnant pools, it frequently takes its seat on the large leaves of water-lilies and other aquatic plants ; being able, by curious discs on its toes, to crawl easily over their smooth surfaces.

It is among the smallest of its tribe, measuring only one and a half inches in length. It is of a greenish-brown, variegated with streaks of green and white, the under surface being of a yellowish-gray, tinged with pink, and the legs banded. Its body is slender, with the hind-legs very long, enabling it to take enormous leaps to escape danger.

THE CHANGEABLE TREE FROG.

Throughout all parts of the continent we shall find a curious little toad, about two inches in length, which possesses the nature of the chameleon—in being able to change its colour according to the tints of the object on which it rests. By this means, so completely does it assimilate its hue to the ground, that it often escapes observation. The changes of colour it thus rapidly passes through are indeed remarkable. From a nearly perfect white, it can assume every intermediate shade to a dark brown. It has a very toad-like look, and possesses skin glands which secrete an acrid fluid. Thus it is able, when attacked, to defend itself, as well as escape observation.

It may frequently be found on old plum-trees, where it climbs in search of the insects which there congregate. We shall frequently hear its voice, especially before rain, for it is a noisy creature. It has a liquid note, sounding like "el" frequently repeated, and then ending with a sharp, short monosyllable.

It leaves its arborial habitation during the breeding season,

TREE FROG.

and makes its way to the nearest pools, where it joins in the concerts of its relatives.

It hibernates during winter, burrowing beneath the damp ground.

THE SPOTTED EFT.

Related to the salamanders, we shall find a curious creature in Pennsylvania, and other parts of the States, known as the spotted eft, or ambystome. It has a thick, convex head, with a rounded muzzle ; and is of a deep violet-black colour above, and purplish-black below, the sides being ornamented with a row of large yellow spots. Unlike other newts, it deposits its eggs in small packets under damp stones. There is another similar creature with mole-like habits, which burrows under the ground, found in various parts of the States.

THE MENOPOMA.

Another of the same order—a formidable and savage creature—is the menopoma, inhabiting the Ohio, Alleghany, and other rivers of the south, frequently, from its propensities, called the young alligator. It is also known as the "ground puppy," the "mud devil," and other well-deserved, if not complimentary names.

It is about two feet in length ; but the teeth, for its size, are small. In appearance, it is ugly in the extreme ; and as, from its voracious habits, it devours a number of fish, and bites fiercely when captured, it is especially hated by the fishermen, who believe it to be venomous, and treat it as seamen do the detested shark.

The above names have been given to it in consequence of its voracity, and its being found generally in muddy bottoms.

THE CONGO SNAKE.

In digging into the mud, sometimes a number of snake-like creatures, between two and three feet long, are turned up—

which have hidden themselves away, often three feet below the surface—in the Southern States. On examination, however, they will be found to have legs, though small and feeble, with only two toes on each foot. They are of a blackish-gray above, and a lighter hue beneath.

Another species of congo snake is found with three toes,—hence the name of three-toed congo snake is given to it.

THE NECTURUS.

Related to the curious eyeless proteus, found in the celebrated cavern of Adelsberg, is an animal very much larger, called the necturus, inhabiting the waters of the Mississippi, and several southern lakes. It is a creature nearly three feet in length, with a thick body, and, being designed to live in daylight, possesses eyes. It is between a fish and a reptile, as it is furnished with large, well-tufted gills ; and, at the same time, has four legs, and four toes on each foot, though it is destitute of claws.

It is of an olive-brown colour dotted with black, and a black streak reaching from one end of the body to the termination of the somewhat thick, short tail.

THE SIREN, OR MUD EEL.

Another curious batrachian, the mud eel, is found in Carolina, in marshy situations. Its total length is about three feet. The head is small, as is the eye, while on each side of it are three beautifully plumed gill-tufts. It has no hind-legs ; while the front pair are very small, and do not aid it in moving along the ground. This it does in the wriggling fashion of an eel ; indeed, when discovered in the soft mud in which it delights to live, the creature, at the first glance, would be

taken for an eel. It has many of the habits of that animal,
living on worms and insects ; indeed, it is difficult to say
whether it should be classed with eels or batrachians. It is,
however, a true amphibian, respiring either in the water by
means of branchiæ, or in the air by means of lungs. It
approaches, in the structure of its head, to the salamanders,
though much less so in its general form and proportions.

The curious " axolotl," which we shall meet with in Mexico,
belongs to a closely allied genus.

GRASSHOPPERS, OR LOCUSTS.

When travelling across the prairies, we may, at times, when
gazing upwards at the sky, see what appears to be a vast
cloud approaching from the horizon. It is produced by
infinite swarms of locusts, or grasshoppers, as they are called
in North America.* About noon they appear to lessen perceptibly the rays of the sun. The whole horizon wears an
unearthly ashy hue, from the light reflected by their transparent wings. The air is filled as with flakes of snow. The
clouds of insects, forming a dense body, cast a glimmering,
silvery light from altitudes varying from 500 to 1000 feet.
The sky, as near the sun as its light will allow us to gaze,
appears continually changing colour, from blue to silvery white,
ashy gray, and lead colour, according to the numbers in the
passing clouds of insects. Opposite to the sun, the prevailing hue is a silvery white, perceptibly flashing. Now, towards
the south, east, and west, it appears to radiate a soft, gray-
tinted light, with a quivering motion. Should the day be
calm, the hum produced by the vibration of so many millions
of wings is quite indescribable, and more resembles the noise

* From Professor Hind's " Red River Exploring Expedition."

popularly termed "a ringing in one's ears," than any other sound. The aspect of the heavens during the period that the greatest flight is passing by is singularly striking. It produces a feeling of uneasiness, amazement, and awe, as if some terrible unforeseen calamity were about to happen.

When the grasshoppers are resting from their long journeys, or in the morning when feeding on the grass and leaves, they rise in clouds as we march through the prairie; and when the wind blows, they become very troublesome, flying with force against our faces, and into the nostrils and eyes of the horses, filling every crevice in the carts. Fortunately, comparatively few take flight on a windy day, otherwise it would be impossible to make headway against such an infinite host in rapid motion before the wind, although composed individually of such insignificant members. The portions of the prairie visited by the grasshoppers wear a curious appearance. The grass may be seen cut uniformly to one inch from the ground. The whole surface is covered with the small, round, green exuviæ of these destructive invaders. They frequently fly at an enormous height above the earth. An engineer engaged in the Nebraska survey, mentions that, when standing on the summit of a peak of the Rocky Mountains, 8500 feet above the level of the plains in Nebraska—being 14,500 feet above that of the sea—he saw them above his head as far as their size rendered them visible.

Grasshoppers are excellent prognosticators of a coming storm. They may be seen at times descending perpendicularly from a great height, like hail—a sign of approaching rain. At this time the air, as far as the eye can penetrate, appears filled with them. Early in the morning they commence their flight, and continue it till late in the afternoon, when they settle round the traveller in countless multitudes, clinging to the leaves of the grass, as if resting after their journey.

They are fearful depredators. Not only do they destroy the husbandman's crops, but so voracious are they, that they will attack every article left even for a few minutes on the ground—saddle-girths, leather bags, and clothing of all descriptions, are devoured without distinction. Mr. Hind says that ten minutes sufficed for them to destroy three pairs of woollen trousers which had been carelessly thrown on the grass. The only way to protect property from these depredators is to pile it on a waggon or cart out of reach.

Two distinct broods of grasshoppers appear—one with wings not yet formed, which has been hatched on the spot ; the other, full-grown invaders from the southern latitudes. They sometimes make their appearance at Red River. However, Mr. Ross, for long a resident in that region, states that from 1819, when the colonists' scanty crops were destroyed by grasshoppers, to 1856, they had not returned in sufficient numbers to commit any material damage. Their ravages, indeed, are not to be compared to those committed by the red locust in Egypt; and yet Egypt has ever been one of the chief granaries of the world.

MEXICO AND CENTRAL AMERICA.

CHAPTER I.

MEXICO.

IF we glance over Mexico, we shall see that the country is, like the continent of which it forms a part, of a triangular shape,—the eastern portion bounded by the Gulf of Mexico, low and flat sandy deserts or noxious marshes being spread over it, and with a narrow belt of level land at the base of the mountains on the Pacific shore. A series of terraces broken by ravines form the sides of a vast table-land,—six thousand feet above the plain,— which stretches from north to south throughout the interior, separated here and there by rocky ridges into smaller plateaux ; while vast mountains in several parts rise from their midst—that of Popocatepetl, the highest in Mexico, reaching to a height of 17,884 feet, with Orizaba, almost of equal elevation, and several mountains not much inferior to them, their snowy summits seen from afar, through the clear atmosphere of that lofty region. Several are active volcanoes, the most curious being that of Jorullo, surrounded by miniature mountains emitting smoke and fire, and presenting the wildest scene of utter desolation. They form pinnacles of the great range of the Andes and the Rocky Mountains. From the midst of the great table-land of Anahuac, flows towards the north the river of Santiago, its course exceeding

VOLCANO OF JORULLO, MEXICO.

four hundred miles, passing in its way through the large lake of Chapala. Some of these table-lands are even eight thousand feet above the sea. The most lofty is so cold, that during the greater part of the day the thermometer varies between 42° and 46°. The great table-land to the east of the Sierra Madre has an elevation which varies from three thousand to six thousand feet. To the west of that sierra,

is the region of Sonora; while eastward, across the Rock
Mountains, is the great valley of New Mexico, watered by th

VEGETATION OF THE TABLE-LANDS OF MEXICO.

Rio Grande del Norte, which has a course of nearly fourtee
hundred miles.

We have thus, in Mexico, a region of elevated plateaux with numerous lofty mountains, steep and broken hill-sides, with deep valleys, watered by numerous streams, and a wide extent of low, level country under the rays of a tropical sun. These several regions possess a great difference in climate, and a corresponding variation in their productions, and, in most instances, in the animals which inhabit them. The domestic animals introduced by the Spaniards, have multiplied greatly, so that vast herds of cattle and horses run wild on the table-lands and lower tracts. Sheep also abound, especially on the northern table-lands. The buffalo makes his way to the great plains bordering the Red River and Arkansas; while deer, in large herds, abound on the higher plains. They are followed, as elsewhere, by packs of wolves and foxes or wild dogs; while the puma makes himself at home here, as he does in Southern America. The bear takes possession of many a mountain cavern; the beaver and otter inhabit the banks of the streams and lakes; the raccoon is found in the woods; and the antelope bounds across the plains.

We know more about the feathered tribes than the mammalia of Mexico. There are upwards of one hundred and fourteen species of land birds, one half of which are unknown in other parts of the world. Still, out of this entire number of species, only one new genus—which connects the family of the tyrant-shrikes with that of the caterpillar-catchers—has been discovered. There are two species of this genus, in both of which the males differ greatly from the females. In this intermediate region we find numerous genera which exist both in Northern and Southern America intermixed. Several South American birds have found their way into Mexico,—as the mot-mots and trogons, the harpy and carracara eagles, the hang-nest, the true and red tanagers, parrots, parrakeets, macaws, creepers, crest-finches, and the fork-tailed and even-tailed humming-birds. Of the genera peculiar to North America,—but which are unknown in the South,—found in Mexico, are the fantailed wagtails, titmice, and worm-eating warblers—blue robins, groundfinch and sandfinch, crescent-starlings and ground-woodpecker. The sandfinch is, however, found in the Brazils. Vast numbers of aquatic birds frequent the lakes and marshes of the table-lands of the interior, as well as the rivers and shores of the coast, nearly the whole of which are well known in the United States, the greater number also inhabiting the Arctic regions.

Among the reptiles, there is one curious creature, peculiar to the country, allied to the siren of Carolina. It is the axolotl, which partakes of the form of a fish, and abounds in many of the lakes in Mexico. It is much esteemed as an article of food by the inhabitants of the neighbourhood.

We cannot speak of Mexico without having our minds drawn to the time of the Aztec monarchy,—when sumptuous palaces, enormous temples, fortresses, and other public edifices covered the face of the country. In the midst of the territory, on the western shore of the large lake of Tezcuco, stood the city of Tenochtitlan, the superb capital of the unfortunate Montezuma, on the site of which has arisen the modern Mexico. Though its glory has long passed away, the enormous ruins which still remain attest its past grandeur. Vast pyramids, on a scale and of a massiveness which vie with those of Egypt, still rear their lofty heads in great numbers throughout the country; while the ruins of other buildings prove that the architecture of Mexico in many points resembled that on the banks of the Nile. Some of these pyramids might rather be called towers. They consist of a series of truncated pyramids placed one above another, each successive one being smaller than the one on which it immediately rests —thus standing in reality upon a platform or terrace. The great pyramidal tower of Cholula is of this character, resembling somewhat the temple of Belus, according to the description given of it by Herodotus. It reaches a height of 177 feet, and the length of each side of its base is 1440 feet. In its neighbourhood are two other pyramids—teocalles, as they are called—of smaller dimensions. These temples, or teocalles, were very numerous, and in each of the principal cities there were several hundreds of them. The top, on which was a broad area, was reached by a flight of steps. On this area were one or two towers forty or fifty feet high, in which stood the images of the presiding deities. In front of the towers was the stone of sacrifice, and two lofty altars, on which fires were kept burning, inextinguishable as those in the temple of Vesta. In the great temple of Mexico there were said to be six hundred of these altars, the fires from which illuminated the streets through the darkest night.

Deeply interesting as is the subject of the architecture and the remarkable state of civilization of the Aztecs, we must not dwell longer upon it, except to mention the cyclopean roads and bridges, constructed of huge blocks of stone, and carried on a continuous level, across valleys, which still remain. There are also, in various parts of the country, excavations, rock-hewn halls, and caverns, generally dome-shaped, the centre apartment lighted through an aperture in the vault. They somewhat resemble the cyclopean fabric near Argos, called the Treasury of Atreus. Not only the buildings, but the hieroglyphics, of the Aztecs, so closely resemble those of the Egyptians, that there appears every reason to suppose they were derived from the same source.

NATURAL BRIDGE IN THE VALLEY OF ICONONZO.

Among the natural curiosities of Mexico, one of the most remarkable is that of the rock-bridge in the valley Icononzo, which might, from its form—until closely examined—be mistaken for a work of art.

The great mass of the population of Mexico consists of the descendants of those tribes which inhabited the country at the time of the Spanish invasion. The language most extensively spoken, as well by the civilized as the savage tribes, is still that of the Aztecs. The people of pure European blood are supposed not to amount to thirty thousand. About a quarter of the population consists of Creoles, descendants of Europeans and Indians known as Mestizos, while there is a small number of Mulattoes, and another race, the Zambos—descendants of Africans and Indians.

Mexico has long been in a chronic state of revolution. From a province of Spain it became an independent empire; afterwards a republic; and once more, under the unfortunate Maximilian, it was placed under imperial rule, finally to fall into a far greater state of anarchy than before.

Before we quit Mexico, a remarkable result of hydraulic action must be mentioned, found on the sea-coast of that region. It is known as the buffadero. At the termination of a long rugged point, the water of the ocean, forced by a current or the waves, is projected through a fissure or natural tube in the rock, forming a beautiful *jet d'eau* many feet in height.

BIRDS OF MEXICO :—THE SCARLET TANAGER.

Among the winter inhabitants of Mexico, one of the handsomest is the scarlet tanager—a small bird, being only six or seven inches in length. It migrates north in the spring, generally making its appearance in the United States about the end of April, where it remains till the breeding season is over.

The colour of the male bird is a brilliant scarlet, with the exception of the tail and wings, which are deep black. The tail is forked, and has a white tip. This gay plumage is, however, only donned during the summer, for when it returns to Mexico in the autumn, its body is covered with a number of grayish-yellow feathers, giving it a mottled appearance. Its note is powerful, but not particularly musical.

Wilson describes it as a remarkably affectionate bird. Having captured a young one, it was placed in a cage high up on a tree. The father bird discovered it, and was seen to bring it food, roosting at night on a neighbouring bough. After

THE BUFFADERO.

continuing to do so for three or four days, he showed by his actions and voice that he was trying to make the young one come out and follow him.   So distressed did he appear, that at last the kind-hearted naturalist set the prisoner at liberty, when it flew off with its parent, who, with notes of exultation, accompanied its flight to the woods.

THE ANIS, OR SAVANNAH BLACKBIRD.

The farmers of Mexico and the Southern States of America whose fields are frequented by the anis, are much indebted to that handsome and somewhat conspicuous bird.   It is of a black hue glossed with green, equalling a pigeon in size—its long tail adding to its apparent length.   Its chief food consists of grasshoppers, locusts, and small lizards, but it rids cows of the ticks and other parasitic insects which fasten on their backs, where they cannot be rubbed off.   So conscious are the cattle of the service thus rendered them, that they will lie down to allow the blackbird to perform the operation at its ease.   It is even asserted that, should the cow neglect to place herself in a suitable attitude, the blackbird will hop about in front of her nose, and allow her no peace till she does as required.

Large flocks of these birds appear together, uttering deafening cries.   When fired at, even though many of them are killed, the survivors hover to a short distance, regardless of the danger in which they are placed.   They build remarkably large nests ; sometimes, indeed, several pairs of birds build one together—much in the same way as do the sociable weaver-birds of Africa—where they live together on friendly terms.

It resembles another African bird in its habit of picking

off ticks from the backs of oxen, the same duty being performed by the South American goatsucker.

MASSINA'S TROGON—THE MEXICAN TROGON.

These birds are remarkable for their beautiful plumage. The first measures about fourteen inches in length. The crown of the head, back, and chest are of a deep, rich green ; the ear-coverts and throat, glossy black; the breast and abdomen, of a rich scarlet. A gray tint covers the centre of the wings, which are pencilled with jet-black lines. The quill-feathers are also black, each being edged with white; and the bill is a light yellow. The females differ considerably from the males. They are shy and retiring birds, and their habits, consequently, are difficult to study.

The Mexican trogon is much smaller than the former, being only a foot in total length, of which the tail occupies nearly eight inches. Few birds are more beautifully adorned than the male trogon. The head is of a bright yellow ; the upper surface of the body, with the chest, being of a rich, glossy green ; while the whole under surface is a bright scarlet. The throat and ear-coverts are black, and a white band of a crescent shape surrounds the throat. The wings are nearly entirely black. The tail is partly black, the two central feathers being green, tipped with black. The females and young males differ greatly, but their plumage is still very handsome.

THE RESPLENDENT TROGON.

The resplendent trogon is a native of Mexico, and, like all its congeners, is fond of hiding its beauty in the dark glades of the rich tropical forests. Its skin is remarkably delicate,

1 MEXICAN TROGON. 2 RESPLENDENT TROGON.

and so thin that it has been compared to wet blotting-paper; while the plumage is so lightly set, that when the bird is shot, the feathers will fall freely from their sockets, through the force of the blow.

The colour of the adult male bird is a rich golden green, on the crest, head, neck, throat, chest, and shoulder-plumes. The breast and under parts shine with as bright a scarlet as the uniform of an English guards-man; the central feathers of the tail are black, and the exterior white, with black bars. The resplendent plumes which overhang the tail are seldom less than three feet in length, so that the total length of this gorgeous bird will frequently reach four feet. The bill is of a light yellow.

This species of trogon feeds chiefly on vegetable diet. We may add that in old times its long plumes were among the insignia of Mexican monarchy, and none but members of the " blood royal " were permitted to wear its gorgeous feathers.

REPTILES :—THE RHINOPHRYNE.

The tongues of frogs, instead of pointing outwards, are directed towards the throat. This species differs from the rest of its tribe, by having its tongue free and pointing forwards. Its rounded head sinks completely into the body, the muzzle being abruptly truncated, so as to form a circular disc in front. So extremely small is the gape, that it would not be supposed, if separated from the body, to have belonged to a frog. On each side of the neck there is a gland, deeply sunk, and almost concealed by the skin.

The body of this curious creature is extremely short and thick, and its feet are half webbed. At the end of each of the hinder feet is a flat, oval, horny spur—its only means of offence and defence, as it possesses no teeth in its head.

It is of a slaty-gray colour, with yellow spots on the sides and back. Occasionally the latter unite, so as to form a jagged line along the back.

THE AXOLOTL.

Among the batrachians found in Mexico is the curious axolotl, which frequents the great lake on which the chief city is built, as well as numerous other lakes, some at a considerable elevation above the ocean. It is between eight and ten inches long, of rather a dark grayish-brown colour, thickly covered with black spots. Those who have seen a newt in its larva state, may form a correct idea of the gills which project from either side of the head.

Naturalists differ in opinion as to whether it is really an

adult batrachian, or merely the larva of some much larger creature. In many localities it is very plentiful ; and the flesh being eatable and of a delicate character, the creature is sold in great numbers in the markets.

THE AXOLOTL.

Being furnished with both kinds of respiratory organs, it can breathe equally well on land or in the water. It has a broad, flat head, blunt nose, and eyes situated near the muzzle. Though living so much in the water, its toes are not connected by intermediate membranes—indeed, they appear only to be intended for service on shore—its tail, nearly as long as its body, serving as a propeller in the water.

CHAPTER II.

CENTRAL AMERICA.

LEAVING the continent of North America, which may be said to terminate at the southern end of Mexico, we enter that extremely irregular portion of land which, now widening, now narrowing again, stretches in a south-easterly direction till it unites with the southern half of the American continent at the Isthmus of Panama. We find in Central America three marked centres of elevation. The first we reach is the great plain, nearly 6000 feet above the level of the sea, on which the city of Guatemala is situated. Numerous volcanic peaks rise from its midst; from it also flow several large rivers, some falling into the Gulf of Mexico, others eastward into the Gulf of Honduras, while smaller streams send their waters westward into the Pacific Ocean. The banks of these rivers are mostly covered with the richest tropical vegetation—the scenery of the river Polochie in Guatemala being especially beautiful. Another high plain occupies the centre of Honduras, and extends into the northern part of Nicaragua. From it also rise numberless streams, some emptying themselves into the Caribbean Sea, and others into the Lakes of Nicaragua and Managua. Further south rises the

volcano of Cartago. Here the Cordilleras ▆▆▆▆▆▆▆▆▆▆▆
character of a vast mountain barrier, but once ▆▆▆▆▆▆▆▆
into low ridges as the chain passes through the Isthmus of
Panama.

As in South America, the Cordilleras run close ▆▆▆▆▆▆
Pacific coast. In consequence, the rivers which ▆▆▆ ▆▆▆▆
their heights have a long course on the Atlantic side, and
have carried down a large quantity of alluvial soil. Here,
too, rain falls in greater or lesser quantities ▆▆▆▆▆▆▆ the
year. The vegetation is consequently rank, and the climate
damp, and proportionately unhealthy. As the trade-winds
blow from the north-east, the moisture with which they are
saturated is condensed against the mountain-sides, and flows
backwards towards the Atlantic. The Pacific slope is, therefore, comparatively dry and salubrious—as indeed are also the
elevated table-lands of the interior.

The whole region is subject to earthquakes, and numberless
volcanoes rise in all directions. In the low ridge which
separates the Lake of Nicaragua from the Pacific are several
volcanic hills, most of them active ; while further to the
north-west, in the district of Conchagua—scarcely more than
one hundred and eighty miles in length—there are upwards
of twenty volcanoes. The two most lofty are found in the
Guatemala range—that of Fuego being upwards of 12,000
feet in height, and that of Agua, 18,000 feet.

Many parts of the interior of the country have been
but very partially explored, and are, indeed, almost unknown.
Of the purely native tribes, most of them have become
mingled with Spaniards or negroes. Parts of the coast are
inhabited by mixed races of Caribs, who have migrated from
St. Vincent, one of the Leeward islands. These Caribs are

RIVER POLOCHIC, GUATEMALA, CENTRAL AMERICA.

known as the Black and Yellow Caribs—the former being the descendants of the survivors of the cargo of an African slaver, wrecked in the neighbourhood of that island. The descendants of the Spaniards are the dominant race, and they have divided the country into various republics, though the greater portion is still in almost as savage a condition as when first discovered.

HONDURAS AND THE MOSQUITO COUNTRY.

The English have, however, a settlement in Honduras; and there is an Indian state forming the eastern portion of Nicaragua, under the government—if so it can be called—of a native king. His territory is known as the Mosquito Country, from the name of the chief native tribe over which he rules.

The climate is very similar to that of the West Indies. On the lower lands a variety of tropical productions can be brought to perfection, while in the high regions cereals of various sorts are abundantly produced.

FAUNA.

The fauna partakes partly of the character of that of the equatorial regions of South America, and of the semi-tropical districts of Mexico. There are several varieties of ant-eaters, similar to those found in the valley of the Amazon, while the gray squirrel of more northern latitudes skips playfully amid the forests of the interior. In the woods and wide savannahs are two or more varieties of deer—one resembling the European deer in colour, but of less size, and adorned with large antlers. The other is of a lighter and browner tint, possessing short, smooth-pointed horns. The peccary is common in the valleys and low ground along the coast; while the waree, or wild hog, runs in large droves in many districts. The tapir, similar to that of the southern continent, also frequents the sea-shore and banks of the rivers; and another species, peculiar to the region, is said to have been discovered lately. There are numerous varieties of monkeys, among which are the brown, the horned, and the little, playful capuchin. The raccoon, as elsewhere, is common, and is noted for its thieving propensities. It lives chiefly on animal food. There is an interesting little opossum of about ten feet in length, of a gray colour, with a somewhat large head, and a long and very flexible tail—the feet being provided with sharp claws. When the young leave the mother's pouch, she can place them on her back, to which they cling, while she scrambles amid the forest boughs. Besides the great anteater, there is the smaller striped ant-eater, and the little ant eater. There is a curious creature, called the quash, resembling the ichneumon, which possesses a peculiarly fetid smell, and is known for its powerful, lacerating teeth. There are several species, also, of the armadillo, distinguished as the three-banded, eight-banded, and nine-banded. The paca is also very plentiful, and becomes easily domesticated. It reaches two feet in length, and its thick, clumsy form, of a dusky brown colour, may be seen scampering through the woods. The agouti, or Indian cony, or rabbit, frequents the same region as the paca, and is about the size of an ordinary hare. It does not, however, run in the same way, but moves by frequent leaps. The jaguar ranges through the whole of this part of the continent, and is remarkable for its large size and great strength. Not only does it frequently kill full-grown cattle, and drag them to its lair far away in the woods, but, if irritated, it does not hesitate to attack human beings.

The tiger-cat, or ocelot, which much resembles a common cat, but is considerably larger, is also found in the forest; but at the sight of man it takes to flight, and is, therefore, less frequently seen than its fiercer relatives. The puma also makes its way from one end of the country to the other; but though destructive to cattle, it is said here, as elsewhere, to fly from

THE JAGUAR OF CENTRAL AMERICA.

the face of man. The savage wolf, the cayote, is frequently met with.

A considerable number of the birds of South America, or of allied species, are found in many parts of the country. This is the home of the resplendent trogon, called the quetzal—the imperial bird of the Quiches. It, however, has but a limited range, being found only in the mountains of Merendon in Honduras, and in the department of Quezaltenango in

Guatemala.   There are numerous varieties of the parrot tribe, many of them of the most magnificent description with regard to their colouring,   Here, also, the forests are adorned with the gay plumage of the red and blue macaws, as also by a toucan with a yellow tail.   It is remarkable not only for its bright colour, but for its curious pendent nests, of which frequently fifty are seen hanging together from the branch of a single tree.   Among the birds of prey, the ever-present

turkey - buzzard and other vultures, hawks, owls, and sea-eagles, are common; as is the Mexican jay, the ring-bird, the rice - bird, swallow, and numerous varieties of humming-birds.   Among the water birds are the pelican, the muscovy, and black duck ; the spoon-bill, plover, curlew, teal, darter ; while herons, ibises,

THE TIGER-CAT, OR OCELOT.

and cranes, are found in great numbers on the shores of the lagoons and rivers.   In the interior of the country the splendid Honduras turkey, as well as the curassow, and several varieties of the wood-pigeon and dove, as also the partridge, quail, and snipe, exist in abundance.

Of the reptile tribes, alligators of great size are found in nearly all the lagoons and rivers.   There is an infinite variety of lizards,—the most noted of which is the iguana, which frequently attains a length of four feet;—and its flesh is here, as in other parts of the continent, esteemed. There are many varieties of serpents, some of which are harmless. Of the venomous species, there are the golden snake, the whip-snake, and the tamagas—the bite of which is considered deadly. So is also that of the corral. It is of the most brilliant colour, covered with alternate rings of green, black, and red. To this last may be added the rattlesnake and the ordinary

THE IMBRICATED TURTLE.

black snake. Most of these snakes are found in the lower region near the sea-coast.

In all the rivers and lakes, tortoises and turtles of several kinds are abundant. The land turtle reaches a foot in length. Its shell is of a dark colour. It is eaten, but is not esteemed of so good a quality as the sea turtle. The coasts are frequented by various species of sea turtle, known as the

THE EDIBLE TURTLE.

green, the hawksbill—which affords the best tortoise-shell to commerce—and the trunk-turtle, which is larger than either of its two relatives. From its flesh is extracted a kind of oil, which is of considerable value.

The hawksbill turtle, which gains that name from its narrow, sharp, and curved beak, like that of a hawk, is also called the imbricated turtle, because its scales overlap each other at their extremities, as tiles are placed on the roofs of houses.

The green or edible turtle is of great size, weighing often six hundred pounds, and being upwards of five feet in length. It gains its name from its rich fat, which is of a green colour ; and its flesh is considered very much superior to that of all its relatives.

The variety and kinds of crustacea are almost numberless, from the largest lobster to the smallest crab. Two species— the mangrove crab, and the white and black land crab—are found near the mouths of the rivers and in all the lagoons ; while the curious soldier crabs, which seem as much at home in one element as in the other, inhabit in vast numbers the trees which lie rotting half submerged in the water. At certain times they may be found making their way into the interior, to return afterwards to the ocean.

The neighbourhood of the ocean, and the rivers and lakes of the interior, swarm with an endless variety of fish ; while the huge manatee, or sea-cow, is found in most of the rivers.

THE MAHOGANY-TREE.

The most valuable production of the forests of this part of the world is the mahogany-tree of Honduras, well-deserving, from its magnificent foliage and vast size, to be called the king of the forest. It is remarkably slow of growth, its increase during half a century being scarcely perceptible.

The life of the mahogany-cutter is wild in the extreme, yet he carries on his occupation in a systematic manner.

Parties, or gangs, are formed, consisting of fifty men, with a captain, or hunter, attached to each. The business of the hunter is to search out the mahogany-trees fit for cutting. To do this, he makes his way through the thick forest to the highest ground in the neighbourhood he can find, and then climbs one of the tallest trees. From thence he surveys the surrounding country in search of the foliage, which presents a yellow, reddish hue, assumed by the mahogany-tree at that season of the year—about August. Having thus discovered a spot on which a number of the sought-for trees grow, he descends, and as rapidly as possible leads his party to it, lest any others on the search should be before them. Huts are now built, roofed with long grass, or the branches of the thatch-palm. His furniture consists of a hammock swung between two posts, and a couple of stones on which his kettle is supported. Stages, on which the axemen stand, are erected round the trees, which are cut down about ten or twelve feet from the ground. The trunk is considered most valuable, on

FLOWERS AND FOLIAGE OF THE MAHOGANY-TREE.

account of the size of the wood it furnishes; but the branches are also of value, from their grain being closer and more variegated.

While one party is employed in cutting down the trees, another is engaged in forming a main road to the nearest river, with others from the various spots where the axemen are at work leading to it. This operation is concluded by the end of December. The trees are now sawn into logs of various lengths, and are squared by the axe, in order to lessen their weight, and to prevent them from rolling in the truck. When the dry weather sets in—about April or May—trucking commences. The trucks are drawn by seven pair of oxen. Each is accompanied by two drivers, sixteen men to cut food for the cattle, and twelve to load the trucks. In consequence of the hot sun during the day, trucking is always carried on at night. A wild scene is presented while the trucks are moving from the forest, each accompanied by several men carrying torches, the drivers cracking their whips and uttering their shouts. Thus they go on till they reach the river's brink, when the logs—each marked with the owner's initials —are thrown into the water, and the trucks return for a fresh load. When the rains commence, the roads are impassable, and all trucking ceases.

As the rivers are swelled by the rains, the mahogany-logs are floated away, followed by the gangs in flat-bottomed canoes, called pit-pans. Their crews are employed in liberating the logs from the branches of the overhanging trees and other impediments, till they are stopped by a beam placed near the mouth of the river. The logs of each owner are now collected into large rafts, in which state they are floated down to the wharves of the proprietors. Here they are newly smoothed, and made ready for shipping to England.

Many other valuable woods come from this region. Rosewood is common on the northern coast of Honduras. The bushes which produce gum-arabic abound in all the open savannahs on the Pacific slope. In the forest is found the

THE IPECACUANHA.

copaiba-tree, producing a healing liquid. Here also are found the copal-tree, the palma-christi, the ipecacuanha—the root of which is so extensively used in medicine—the liquid amber, as well as caoutchouc. Here the vast ceiba, or silk-cotton tree, is abundant, from which canoes are frequently hollowed out. Indeed, a considerable number of the trees found on the banks of the Orinoco and Amazon here also come to perfection.

HUMMING-BIRDS.

HUMMING-BIRDS :—THE SLENDER SHEAR-TAIL.

Central America is the home of several beautiful species of those minute members of the feathered tribe—the humming-birds. Among them is found the slender shear-tail, which will be known by its deeply-forked black tail, its wings of purple-brown, and its body of deep shining green, changing to brown on the head, and bronze on the back and wing-coverts. The chin is black, with a green gloss; the throat is of a deep metallic purple; while a large crescent-shaped mark of buff appears on the upper part of the chest. There is a gray spot in the centre of the abdomen, and a buff one on each flank, the under tail-coverts being of a greenish hue.

The female differs greatly from her consort. Her tail is short, the central feathers being of a golden green; the exterior ones rusty red at their base, and black for the greater part of their length, with white tips. The upper part of her body is also of a golden green; the lower of a reddish-buff.

THE RUFUS FLAME-BEARER.

The beautiful little rufus flame-bearers belong to the genus Phæthornis. They are known by their long, graduated tails, all the feathers of which are pinnated—the two central ones extending far beyond the others. "They may be seen early in the year, darting, buzzing, and squeaking in the usual manner of their tribe, engaged in collecting sweets in all the energy of life, appearing like breathing gems—magic carbuncles of glowing fire—stretching out their glorious ruffs, as if to emulate the sun itself in splendour. The female sits towards the close of May, when the males are uncommonly quarrelsome and vigilant, darting out as the stranger approaches the nest, looking like angry coals of brilliant fire, returning several times to the attack with the utmost velocity, at the same time uttering a curious, reverberating, sharp bleat, somewhat similar to the quivering twang of a dead twig, and curiously like the real bleat of some small quadruped. At other times the males may be seen darting high up in the air, and whirling about each other in great anger and with much velocity.

"The nests are funnel-shaped, measuring about two and a quarter inches in depth, and one and three-quarters in breadth at the upper part, composed of mosses, lichens, and feathers woven together with vegetable fibres, and lined with soft cotton."

This description is given by Mr. Nuttal the naturalist, and quoted by Audubon.

PRINCESS HELENA'S COQUETTE.

This beautiful little gem—a native of Vera Paz, in Guate-

mala—is adorned somewhat after the fashion of the Birds of
Paradise, its head being ornamented with six long, green,
hair-like feathers, three on either side of the body.   The

1. PRINCESS HELENA'S COQUETTE.   2. TUFTED COQUETTE

upper part is of a coppery bronze colour, a band of buff
crossing the lower end of the back.   The face is green; and
the throat is adorned with emerald feathers surrounded with

others long and white.   These start from the neck, being
edged with blue-black.   Beautifully adorned as is the male,
the hen bird possesses neither crest nor neck-plumes, her
colour being of a dull, bronze green, and grayish - white
sprinkled with green on the under part of the body.

THE SPARKLING-TAIL HUMMING-BIRD.

The little sparkling-tail is one of the boldest and most
familiar of its tribe, being seen flitting from flower to flower
among the gardens in Guatemala, and remaining with perfect
confidence even while people are moving about near it.   It
is one of the smallest of its tribe—the nest being also of
a proportionate size, formed of various delicate fibres, such as
spider's webs and cottony down, and covered with lichens.
Within it the female lays two eggs, scarcely larger than peas,
of a delicate, almost transparent, pearly white.   This nest is
secured to a slight twig by spider's webs.

The general colour of the male is bronzed green above,
with a crescent-shaped white mark on the lower part of the
back.   It has a rich metallic blue throat, changing in certain
lights ; and the wings are of a dark purple-brown.   The tail
is composed of feathers of different tints—the two central of
a rich, shining green ; the next, green, marked with bronze ;
and the outer, dark brown, with triangular white spots on the
inner web.

The whole length of the bird, with its forked tail, is about
four inches.   The hen has a shorter tail, the feathers purple-
black, bronzed at the base, and most of them tipped with
white and ringed with buff.   The upper part of the body is
of a rich bronzed green ; and the lower, a rusty red.

Many other beautiful humming-birds appear throughout dif-

LOCUSTS.

ferent parts of Mexico and Central America; but we may grow weary even when examining caskets of the most brilliant gems; and we shall have many others to describe when we reach the southern part of the continent.

LOCUSTS.

Insect life is as active in Central America as in other parts of the tropics. The most dreaded insect is the locust, which makes periodical attacks on the plantations, and in a single hour the largest fields of maize are stripped of their leaves, the stems alone being left to show that they once existed. This creature is called by the natives the "chapulin," or langosta. They make their first appearance as little wingless things, swarming over the ground like ants, when they are called "santones." In order to destroy them, the natives dig long trenches, into which they are driven, when, unable to leap out, they are easily buried and destroyed. Still, vast numbers escape, when they appear in enormous columns, darkening the air, and as they sweep onwards, destroy every green thing in their course. They cover the ground on every side, then rising in clouds, fill the atmosphere with their multitudes, causing the trees to appear brown, as if seared by fire. Frequently, as their hosts sweep onwards, they are seen falling like flakes in a snow-storm from a dark cloud. Every device that the farmer can think of is employed to prevent their settling : sulphur is burned, drums beaten, guns fired, and other noises made. Often, by such means, a plantation is preserved from destruction ; but when the columns once alight, no device avails to save the plantation from speedy desolation.

This locust or grasshopper is generally from two and a half to four inches in length, but specimens sometimes appear five inches long ; and it may be conceived what an enormous amount of food such monsters must consume.

CHAPTER III

RUINS OF CENTRAL AMERICA.

IN all parts of Central America are found numerous signs that the country was, in bygone days, inhabited by a numerous population far more advanced in civilization than the tribes which peopled it when first discovered by Columbus and his companions. In Yucatan and Chiapas, especially, ruins of numerous houses exist, with elaborately carved monuments and large buildings, bearing a remarkable resemblance to those of Egypt and Babylon. Throughout Nicaragua and other districts many remains—such as tombs, monuments, and edifices—are found, as well as carved rocks, which were probably the work of a people of still greater antiquity than those who inhabited the first mentioned region.

Dr. Seeman describes some rocks near the town of David, in Chiriqui, on which characters are engraved similar, or indeed absolutely identical, with inscriptions which have been found in the northern parts of the British Islands. The rock is fifteen feet high, nearly fifty feet in circumference, and rather flat on the top. Every part—especially the eastern portion—is covered with incised characters about an inch or half an inch deep. The first figure on the left hand side represents the radiant sun, followed by a series of heads with some variation. These heads show a certain resemblance to one of the most curious characters found on the British rocks. They are followed by scorpion-like and other fantastic figures. The tops of the stones on either side are covered with a number of concentric rings and ovals, crossed with lines. He considers them to be symbols full of meaning, and recording ideas held to be of vital importance to the people who used them, and whose names have become a matter of doubt.

In the district of Chontales, a vast number of ancient tombs are met with in almost every direction. They are found in plains having a good drainage, such as was generally selected by the Indians for the sites of their villages. These tombs are of different heights and sizes. Some are about twenty feet long by twelve feet wide, and eight feet above the ground. In one which was opened was found a round pillar seven feet high and eighteen inches across, which was standing upright in the centre of the tomb. There was a hand-mill for grinding corn—in shape like those still in use in the country—a knife ten inches long, a hatchet like a reaping-hook, and a tiger's head of natural size,—all of stone. In some instances gold ornaments have been found, but not in sufficient numbers to induce the people to destroy the relics.

The Indians inhabiting Nicaragua in ancient days did not apparently construct any large temples or stone buildings, as some other natives of Central America have done. They, however, formed stone figures of considerable size, which remind us greatly of those which exist in Easter Island in the Pacific. These stone figures, often of colossal dimensions, are

of two different descriptions—the one having a mild, inoffen-
sive expression of countenance ; while the others, presenting a
combination of both human and animal, have invariably a
wild, savage look, apparently for the purpose of terrifying the
beholders.   The first, it is supposed, are the idols which the
ancient Nicaraguans worshipped before the Aztec conquest of
their country ; while the latter were introduced when the
people had been taught to engage in the bloody rites practised by the Mexicans.

These stone monuments, though similar, as has been re-
marked, to those of Easter Island, and to others found far
away across the Pacific, are strong corroborative proofs that
America was first peopled by tribes who made their way by
various stages from the continent of Asia, though, at the same
time, that long ages have passed away since they first left
that far-distant region—the cradle of the human race.   The
Indian priests, like the Druids of old, appear to have chosen
the hill-tops and mountain-sides, shady groves and dark
ravines, for the sites of their temples or places of worship.

From the midst of Lake Managua, in Nicaragua, rises the
volcanic island of Momotombita, towering in a perfect cone
towards the blue sky.   In the midst of a natural amphitheatre
on the slope of the mountain were discovered a large number
of statues (fifty or more), arranged in the form of a square,
their faces looking inwards.   Many were cast down, but
others stood erect, though all apparently had been more or
less purposely mutilated.   Some of the figures represent
males, but others are undoubtedly those of females.   They
are cut in black basalt of intense hardness.   The features of
the face of one, which has been conveyed to the Museum at
Washington, are singularly bold and severe in outline.   The

brow is broad, the nose aquiline, while the arms and legs are rudely indicated. Other curious idols have been dug up in the neighbourhood of the town of Leon. The Spanish priests, anxious to put down the ancient idolatry from the time of their arrival in the country, have taken pains to destroy these idols, and many have been mutilated and others buried by their orders.

In the island of Zapetero, rising out of Lake Nicaragua, there are a still greater number of statues—some from eight to twelve feet in height, and others of still greater magnitude —elaborately carved out of hard stone. Sometimes they are placed round mounds which have evidently served the purpose of altars, on which human sacrifices probably were offered. One of the most interesting which has been brought to light is twelve feet high, sculptured from a single block, and representing a human figure seated on a high pedestal, the stone at the back of the head being cut in the form of a cross. The limbs are heavy, and the face large and expressive of great complacency.

Some of the idols represent an animal, apparently a tiger, springing upon the head and back of a human figure. One —also at the Washington Museum—represents a man squatted on his haunches, with one hand at his side, and the other placed on his breast. The head is erect, and the forehead encircled by a fillet, much carved. The features are unlike most others—indeed, it seems as if each one had its individual characteristic. A jaguar appears on the back of this statue, its fore-paws resting upon the shoulders, and its hind ones upon the hips, while it grasps in its mouth the back part of the head of the figure.

Although many of the figures represent human beings,

others are those of animals.   One, a jaguar, is seated on its haunches, the head thrown forward, the mouth open,—the attitude and expression being that of great ferocity.   It is very boldly sculptured.   Another,. a very well proportioned human figure, is seated on a square throne raised five feet from the ground.   It is remarkable for having on its head another monstrous head, representing some fierce animal.   The heads of several of the idols are thus surmounted.   These symbolical heads were probably introduced with the same object as those which were so general among the Egyptian idols.

In the midst of this collection of idols are two or more oblong stones, on the sides of which are hieroglyphical inscriptions.   In the centre are hollow places, probably designed to receive the blood of the victims.

It is remarkable that the heads of many of the figures are surmounted with cross-shaped ornaments similar to the one discovered at Palenque by Mr. Stevens.   One of these crosses —which no doubt had their origin in Babylon, where they are well-known symbols—was set up by the Spaniards in the convent-church of Tonala, and there venerated.

The Mexicans possessed a symbol called the *Tonacaquahutl*, or "tree of life," which was represented with branches somewhat in the form of a cross, surmounted by a bird.   This symbol also appears on a tablet discovered by Mr. Stevens at Palenque.   In various parts of the country terra cotta figures have been dug up.   Some of them are rude, but others are extremely artistic; and though not equally graceful, resemble much, in the form of the limbs, many Egyptian figures. Among them is a figure from the island of Ometepe, which represents an alligator upon the back of a human figure, which apparently originally surmounted a large vase.

Mounds similar to those found in the valley of the Mississippi have been discovered in Honduras. But by far the most interesting remains are those of Palenque, in Chiapas; of Copan, in Honduras; and of Uxmal and Chi-chen, in Yucatan. Here are extensive ruins of cities, containing the remains of pyramids, and the walls of massive buildings, broken columns, altars, statues, and numberless sculptured fragments, showing that a large population inhabited this country, and that the people had attained a considerable knowledge of the arts, though, at the same time, they seem to have been sunk in the grossest idolatry.

In the western part of Honduras, adjoining the province of Guatemala, are extensive ruins, which stretch for more than two miles along the banks of the river Copan. The outer walls, which run north and south along the margin of the stream, are from sixty to ninety feet high; while other walls, of a similar character, surround the principal ruins. Within these walls are extensive terraces and pyramidal buildings, massive stone columns, idols, and altars covered with sculpture. The numerous terraces and pyramids are also walled with cut stone, and ornamented with carved heads of gigantic proportions, and colossal idols of solid stone from ten to fifteen feet in height. The altars in front of the statues are of single blocks of stone, many of them richly carved, but all differing from each other. One of the most remarkable altars stands on four globes cut out of the same stone. It is six feet square and four feet high, its top covered with hieroglyphics, while each side represents four individuals. The figure is sitting cross-legged, in the Oriental fashion, and the head-dresses are remarkable for their curious and complicated forms. All have breastplates, and each holds some article in his hand.

From these carvings we read, though indistinctly, some of the characteristics of the people. From the absence of all weapons of war, however, wo may suppose them peaceable, though grossly idolatrous, and, from being unwarlike, easily subdued.

On entering the town, after some adventures, Mr. Stevens made his way to an area, which he ascertained to be a square, with steps on all sides, almost as perfect as those of the Colosseum. He ascended the steps, which were ornamented with sculptures, till he reached a broad terrace, one hundred feet in height, overlooking the river. The whole terrace was covered with trees, among which were two gigantic cotton-trees of about twenty feet in circumference, extending their roots fifty to one hundred feet round, and which had, in many places, displaced the stones. Among other ornaments were rows of gigantic heads, which, no doubt, were intended to represent those of apes; for amongst the fragments were the remains of the body of a colossal ape, strongly resembling in outline and appearance one of the four monstrous animals which once stood in front of the obelisk of Luxor, and which, under the name of Cynocephali, were worshipped at Thebes. This fragment was about six feet high.

No verbal description can give a correct idea of the elaborate workmanship of the numberless idols. One, described by Mr. Stevens as the most beautiful in Copan, he considers equal to the finest Egyptian sculpture; and thinks, indeed, it would be impossible, with the best instruments of modern times, to cut stones more perfectly. They are generally from twelve to fourteen feet in height, about four wide, and two or three deep. On the front is, in all cases, a human head, with arms and hands, surrounded by the most intricate carving.

Frequently other smaller heads appear below the large one. In many instances the legs and feet, as well as the body, are represented. The backs and sides are covered with the most elaborate hieroglyphics, deeply carved—the whole forming a mass of rich ornamentation. Before several of the idols stand altars, also carved in the same finished way.

The most interesting figure—which, unlike all the others, is remarkable for its simplicity—is that of a human being, bearing on its head a heavy cross-like crown. It cannot fail to remind those acquainted with the idols of Babylon of the Triune God represented in the sculptured stones of those far-famed ruins.

STONE QUARRIES.

Some two or three miles from the ruins are the quarries, from which the stones for the buildings and statues of Copan are evidently taken. Here still exist huge blocks of stone, in different degrees of preparation. Near a river was found a gigantic block, much larger than any in the city, which was probably on its way thither, to be carved and set up, when the labours of the workmen were arrested. It is difficult to conjecture how these vast masses were transported over the irregular and broken surface of the country, and particularly how one of them was set up on the top of a mountain two thousand feet in height.

A place of this name was captured by Hernandes de Chaves at the time when its now broken monuments, ruined terraces, walls, and sculptured figures, were entire, and were all richly painted; and it seems strange that Europeans could have beheld its wonders without spreading the report of them throughout the civilized world, yet no account of this strange city was extant till it was visited by Mr. Stevens.

PALENQUE.

Still more curious and interesting than the last described city, are the ruins of Palenque, in the province of Chiapas, bordering upon Yucatan. One of the chief structures of this ancient city stands on an artificial elevation 40 feet high, 310 feet in length, and 260 feet in width. The sides were originally covered with stones, which have been thrown down by the growth of trees. On the summit are the ruins of a building, known as the Palace, about 25 feet in height, with a front measuring 228 feet by 180 feet deep. In front were, originally, fourteen doorways, with intervening piers, covered with human figures, hieroglyphics, and carved ornaments. The walls are of stone, laid with mortar and sand; and the whole is covered by stucco, nearly as hard as stone, and richly painted. On each side of the steps are gigantic human statues carved in stone, with rich head-dresses and necklaces.

In one of the buildings is a stone tower of three stories, thirty feet square at the base, and rising far above the surrounding walls. The walls are very massive, and the floors are paved with large square stones. In one of the corridors are two large tablets of hieroglyphics.

There are numerous other buildings, all standing on the summits of similar pyramids. In several of the buildings the roofs still remain, and preserve the stuccoed ornamentation with which the walls are adorned. The colours, in many of them, are still bright; and could the hieroglyphics with which they are surrounded be read, they would probably give as clear a history of the departed inhabitants as do those found in the tombs on the banks of the Nile. The most remark-

able figures are the bas-reliefs, in stucco, representing a woman
with a child in her arms—which forcibly remind us of the
statues in ancient Babylon representing the goddess mother and
son (the same worshipped in Egypt under the names of Isis
and Osiris; in India, even to this day, as Isi and Iswara;
and also in China, where Shingmoo, the holy mother, is represented with a child in her arms, and a glory round her head).
It is impossible, looking at these figures, to suppose otherwise
than that they were derived from the same source whence the
idols of Egypt, Greece, and pagan Rome had their origin.

RUINS OF QUICHÉ.

In the north-east of Guatemala are the ruins of another city,
the capital of the province of Quiché.   It is surrounded by
a deep ravine, which forms a natural foss, leaving only two
very narrow roads as entrances, guarded by the castle of
Resguado.   The palace of the kings, which stood in the centre
of the city, surpasses every other edifice, competing in magnificence with that of Montezuma in Mexico.   It was constructed of hewn stones, of various colours.   So large was the
city, that it could send no less than seventy-two thousand
fighting men to oppose the Spaniards.   The whole palace is
now, however, completely destroyed, and the materials have
been carried away to build a village in the neighbourhood.
The most conspicuous portion of the ruins remaining is called
El Sacrificatorio.   It is a quadrangular stone structure, rising
in a pyramidal form to the height of thirty-three feet.   At
the corners are four buttresses of cut stone.   Steps lead up on
the eastern side.   On the top it is evident that an altar was
once placed, for the sacrifice of human victims, which struck
even the Spaniards with horror.   The whole was in full view

of the people who collected round the base.   The ruins differ
entirely from Copan and Palenque.   Here no statues, carved
figures, or hieroglyphics are seen.   It is therefore supposed
that these cities are of a much older date, and built by another
race.

UXMAL.

The most magnificent and perfect remains in the country
are those of Uxmal, about fifty miles south of Merida, the
principal city of Yucatan.   Here, amid the dense forest, are
found walls of considerable elevation, with very extensive
buildings,—the walls still standing to their full height, and
even the roofs, in some places, perfect.   The largest building
—supposed to be the palace of the sovereign—stands on the
uppermost of three terraces, each walled with cut stone.   It
is 322 feet in length, 39 broad, and 24 high.   The front has
thirteen doorways; the centre of which is 8 feet, 6 inches wide,
and 8 feet, 10 inches high.   The upper part is ornamented
with sculpture in great profusion, of rich and curious workmanship.   The walls are covered with cement ; and the floors
are of square stones, smoothly polished, and laid with as
much regularity as that of the best modern masonry.   The
roof forms a triangular arch, constructed with stones overlapping, and covered by a layer of flat stones.   It is remarkable
that the lintels of the doorways are of wood, known as
Sapote wood.   Many of them are still hard and sound, and
in their places ; but others have been perforated by worm-
holes, their decay causing the fall of the walls.

Two other large buildings, facing each other, are embel-
lished with sculpture, the most remarkable features of which
are two colossal serpents, which once extended the whole
length of the walls.   Further on are four great ranges of

edifices, placed on the uppermost of three terraces. The plan of these buildings is quadrangular, with a courtyard in the centre. The walls are, like the others, ornamented with rich and intricate carving, presenting a scene of strange magnificence. One of the buildings is 170 feet long, and is remarkable for the two colossal entwined serpents which run round it, and encompass nearly all the ornaments throughout its whole length. These serpents are sculptured out of small blocks of stone, which are arranged in the wall with great skill and precision. One of the serpents has its monstrous jaws distended ; and within them is a human head, the face of which is distinctly visible in the carving.

The most tastefully ornamented edifice is know as the "House of the Dwarf." It stands on the summit of a lofty mound, faced with stone, nearly ninety feet high, the building itself being seventeen feet high. Its purpose it is difficult to divine.

Scattered throughout the ruins are a number of dome-shaped subterraneous chambers, from eight to ten feet deep, and from twelve to twenty in diameter. The floor is of hard matter, and the walls and ceilings of plaster. A circular hole at the summit of each, barely large enough to admit a man, is the only opening into them. It is not known whether they were used as cisterns, or for granaries, like those of Egypt.

OTHER RUINS.

The whole country to the south of Uxmal is covered with ruins. At a place called Labra, there is a tower richly ornamented, forty feet in height, which stands on the summit of an artificial elevation. In another place there is one forty-five feet high ; along the top of which, standing out from the wall, is a row of deaths' heads—or perhaps monkeys' heads —and underneath are two lines of human figures, greatly mutilated.

At Kewick, a short distance from Labra, are numerous other ruins, mostly remarkable for the simplicity of their architecture and the grandeur of their proportions. It is still uncertain whether these cities were inhabited by the unhappy people conquered by the Spaniards, or whether they were built by a race which, from some unknown cause, had already passed away. We see how completely the Mexicans and Peruvians, after the conquest, sunk from their comparatively high state of civilization into barbarism; and such might have been the case with the inhabitants of these cities. Their origin will probably for ever afford matter for speculation.

The different cities vary in their style of architecture almost as much as as they do from those of Assyria or Egypt; but when we come to examine the sculptures, we may be able to trace a much stronger resemblance. The statues of the woman and child, the cruciform ornaments, the serpents and gigantic heads of apes, as well as those of the typical heads of savage animals surmounting the heads of the statues, are all to be found on the banks of the Nile, and were probably derived from the same central source. While the tribes who proceeded westward peopled Egypt, others, among whom a similar system of idolatry prevailed, may have migrated towards the east, and finally made their way across the Pacific to the shores of America.